ACOUSTICS RESEARCH AND TECHNOLOGY

TINNITUS

CAUSES, TREATMENT AND SHORT AND LONG-TERM HEALTH EFFECTS

ACOUSTICS RESEARCH AND TECHNOLOGY

Additional books in this series can be found on Nova's website under the Series tab.

Additional e-books in this series can be found on Nova's website under the e-book tab.

ACOUSTICS RESEARCH AND TECHNOLOGY

TINNITUS

CAUSES, TREATMENT AND SHORT AND LONG-TERM HEALTH EFFECTS

FRANCESCO SIGNORELLI
AND
FRANCIS TURJMAN
EDITORS

New York

Library of Congress Cataloging-in-Publication Data

ISBN: 978-1-63117-556-5

Library of Congress Control Number: 2014934749

Published by Nova Science Publishers, Inc. † New York

Contents

Preface

Tinnitus is a benign condition, which can nonetheless have upsetting consequences for the patients. Understanding the physiopathologic mechanisms underlying this symptom is crucial for orienting management and treatment. A world-class group of neuroscientists was gathered together to elaborate this book, and they have done a skillful job. Apart from delineating their area of expertise, they have managed to give the reader an accurate hint about what the current implications of studying tinnitus are, both in research and clinically, allowing those involved in the field to stay abreast on the advancements. This book is ideal for researchers and clinicians, and it can be a valuable reference for anyone, including students, residents or seasoned specialists, seeking to keep up to date with the latest developments in both basic and clinical research on this condition.

Key features of this book are:

- four chapters : Pathogenesis of tinnitus: the possible role of oxidative stress; *Tinnitus: Causes, Treatment and Short & Long-Term Health Effects; potential roles of* Na^+ influx and Src family kinases (SFKs) in the development of tinnitus and hearing loss; Mindfulness Based Tinnitus Stress Reduction: A New Treatment With Ancient Roots
- High-yield of pathogenesis principles of tinnitus
- Straightforward explanations of the basic as well as the latest advances in the diagnosis and treatment of tinnitus

Prof. Francesco Signorelli
Associate Professor of Neurosurgery, University Magna Græcia, Catanzaro, Italy
and
Consultant Neurosurgeon, Hospices Civils de Lyon, Hôpital Neurologique et Neurochirurgical, Department of Neurosurgery, Lyon, France

Prof. Francis Turjman
Professor of Neuroradiology, Division of Interventional Neuroradiology, Hôpital Neurologique et Neurochirurgical P. Wertheimer and University Claude Bernard Lyon 1, Lyon, France

In: Tinnitus
Editors: F. Signorelli and F. Turjman
ISBN: 978-1-63117-556-5

Chapter 1

Pathogenesis of Tinnitus: The Possible Role of Oxidative Stress

Andrea Ciorba and Chiara Bianchini
ENT and Audiology Department, University Hospital of Ferrara, Italy

Abstract

Background. Tinnitus is one of the most common disabilities in developed countries; however the comprehension of its generation mechanisms still remains a challenge for pathophysiologists, otolaryngologists and audiologists. The possible role of oxidative stress in its pathogenesis has been claimed by several authors and also the use of antioxidants in its treatment has been promoted.

Methods. A systematic review of the English literature among tinnitus and oxidative stress has been performed through Medline searches.

Results and Conclusions. Oxidative stress could have an implication in the pathogenesis of tinnitus secondary to some inner ear pathologies; nonetheless its role in idiopathic tinnitus is still unclear. On one hand, treatment with antioxidants might be appropriate for tinnitus secondary to cochlear pathology induced by, or linked to oxidative stress; on the other hand, it is still necessary to understand the etiopathologic mechanism underlying the onset of idiopathic tinnitus in order to develop a causal approach.

Introduction

Deafness and tinnitus are between the most common disabilities in developed countries, and there is a considerable social and economic demand for the development of new therapeutic approaches for these conditions [1-3]. The prevalence of tinnitus in adults is estimated between 10% and 15% in developed countries as reported in the current literature [1-3].

It is well known that the loss of hair cells within the human inner ear results in hearing disorders that can significantly impair the quality of life. The human cochlea is unable to replace hair cells (inner and outer) once lost and this is the cause of an irreversible hearing impairment. Tinnitus is a perception of sound in absence of a corresponding external sound. It is reported to be a common symptom, however the comprehension of its generation mechanisms still remains a challenge. Tinnitus may be present as the only symptom and in a situation of normal hearing (idiopathic tinnitus), or it can be associated to sensorineural hearing loss. Particularly in this case, several Authors have already claimed the possible role of oxidative stress in order to explain its pathogenesis [1-3].

The suggestion that antioxidants could be used to prevent or repair a labyrinth damage comes from the studies on ototoxicity [1-3]. Reactive oxygen species (ROS) are oxygen-based radical molecules that may be produced in the human body in both physiological and pathological conditions. In a healthy situation, they are produced as regulatory means, intercellular signals or bactericidal agents. An increased ROS presence may be caused either by overproduction of ROS or from depletion of antioxidant defenses [4], and it is already known that inner ear hair cell damage and/or death may occur as result of ROS-mediated damage [4]. It has been established that ROS are generated in hair cells exposed to several insults including cisplatin, aminoglycosides, or noise, and that antioxidant drugs can reduce the amount of intracellular damages ROS mediated. In highly oxidative conditions, endogenous antioxidant pathways can become overwhelmed, and free oxygen radicals may become abundant. Particularly, aminoglycosides, such as gentamicin, can activate inducible nitric oxide synthase in inner ear tissues, triggering an increase in nitric oxide [3,5]. As result of the increased intracellular free radicals production, apoptosis (active cell death) begins, and G proteins, such as Ras, and GTPases, such as Rac are triggered. These events result in the activation of a family of stress-activated protein kinases, such as mitogen-activated protein kinase (MAPK) and c-Jun N-terminal kinase (JNK). The increased activity of these enzymes is accompanied by increased intracellular Ca2+ concentrations and the release of cytochrome c from mitochondria. Cytochrome c release (usually mediated by Bax, a protein that enhances apoptotic cell death), causes mitochondrial membrane damage mainly by pore formation [3, 5]. The release of cytochrome c from mitochondria activates caspase 3 and/or 8 (caspase dependent pathway); to degrade chromatin, caspase-3 acts on a set of DNAses and nucleases (PARP 1). To degrade the cell membrane and the cytoskeleton, caspase-3 acts on other molecules such as gelsolin and fodrin [6-10]. Also another class of calcium-dependent proteases, so called calpains, can be activated as a consequence of cytochrome c release from mitochondria and can lead to HCs disruption (calpain cascade) [5, 11, 12, 13]. As result of the above mechanisms prompted by ROS, inner ear hair cells can be damaged and then destroyed.

A systematic Pubmed database search up to June 2013 (going back for 10 years) was performed; full text articles were obtained when the title, abstract or key words suggested that the study may be eligible for this review. The search was carried out independently, and restricted to papers in English language. Other papers were also identified from the references in the published literature.

The medical subject heading (MeSH) used included: oxidative stress, inner ear, free radicals and tinnitus.

Oxidative Stress and Tinnitus due to Inner Ear Diseases

Cochlear dysfunction has been implicated in the generation of tinnitus [14]. Particularly, it has been reported that an increase of ROS intracellular level may be responsible for cochlear damage in various pathological conditions which express themselves clinically with hearing loss and tinnitus. Therefore, the identification of possible sources of cochlear ROS and understanding the implementation mechanisms of ROS-mediated cochlear damage, could help in developing new approaches for the prevention and treatment of inner ear disorders.

As presented above, ROS are responsible for direct intracellular lipids, proteins and DNA damages, triggering apoptosis or necrosis in the inner ear [13, 15]. Interestingly, it has been recently observed that the different cellular components of the cochlea, do not share the same vulnerability to the injury induced by ROS, as outer hair cells appear to be more susceptible to ROS damage, and especially those at the base of the cochlea, while supporting cells should have a greater capacity of survival [16, 17]. Moreover, researchers have found that glutathione (an antioxidant agent) is more expressed in the apical hair cells and NOX3, responsible for the production of superoxide, is more expressed in the basal hair cells and neurons of the spiral ganglion [16, 17].

Aminoglycosides. Once inside the cell, aminoglycosides induce the generation of ROS due to the formation of an aminoglycoside-iron complex, which catalyses the oxidation of unsaturated fatty acids located in the inner leaflet of the plasma membrane [18-21]. ROS subsequently activate apoptotic or necrotic intracellular pathways [18, 22]. They promote the opening of the mitochondrial permeability pores and they also activate the JNK pathway leading to hair cell apoptosis [18, 23, 24].

Cisplatin. Some studies have demonstrated the direct cytotoxic mechanisms of cisplatin, such as DNA and mitochondrial damage, with the formation of ROS [18, 25]. Increased intracellular ROS trigger mitochondrial release of cytochrome c, through activation of the pro-apoptotic Bcl-2 family proteins. Subsequent apoptosis is mediated by the activation of pro-caspase-9 and -3 [18, 26].

Noise. Increasing evidence suggests that ROS have an important role in noise-induced cochlear damage: increased levels of superoxide anion, hydroxyl radical are observed in the cochlea after intense noise exposure. In addition, one of the best established endogenous “fingerprints” of ROS action is the peroxidation of polyunsaturated fatty acids. Markers of lipid peroxidation have been demonstrated in the hair cells, supporting cells, spiral ganglion neurons and stria vascularis after noise trauma. Thereafter, ROS accumulation initiates a complex cascade of biochemical processes that includes the activation of JNK, the release of cytochrome c from the mitochondria and the activation of procaspase-8, -9 and -3 (intrinsic pathway of apoptosis) [18, 27-31].

Presbycusis. This is an extremely complex, multifactorial process, implying high frequency hearing loss concomitantly with physical signs of ageing. Indeed, genes that protect against oxidative stress are involved in development of age-related hearing loss. Studies of the aging cochlea showed a decrease of antioxidant defences such as glutathione level in the auditory nerve or antioxidant enzymes in organ of Corti and spiral ganglion neurons [18, 32-34].

Sudden sensorineural hearing loss (SSNHL). Aetiology and pathogenetic mechanisms of SSNHL are still not fully clarified, even if viral infections, autoimmune processes and especially vascular disorders seem to play an important role. Whatever the cause, impaired cochlea perfusion seems to be a key pathogenetic factor of SSNHL because of cochlea terminal vascularisation and its extreme sensitivity to anoxia or hypoxia. Recent studies have stressed the importance of oxidative stress as a risk factor for microvascular damage and hypooxigenation [35]. Thus, imbalance between ROS and total antioxidant capacity is thought to be a potential pathogenetic mechanism leading to cellular damage also in SSNHL [35, 36].

Meniere Disease (MD). Recently Calabrese et al. observed that patients affected by MD are under condition of systemic oxidative stress and that in peripheral blood of MD patients are present measurable increased markers of cellular stress response and oxidative stress factors [37].

Immunomediated sensorineural hearing loss (ISHL). The ISHL is a rare disease accounting for less than 1% of all cases of hearing impairment, tinnitus or dizziness. It is characterized by a rapidly progressive, often fluctuating, bilateral SNHL over a period of weeks to months. Its pathogenesis is still unknown even if it is believed that inflammatory mediators such as endotoxin and free radicals are involved. Responsiveness to steroid is high and with prompt treatment, hearing loss may be reversible [38].

The role of ROS in the pathogenesis of inner ear damage seems to be clear, at least in certain conditions. In this way, our research group provided the first evidence of the presence and production of an oxygen radical species (superoxide) in human inner ear perilymph (perylymph was sampled from subjects affected by profound hearing loss and treated with cochlear implantation) [4].

Moreover, it has been reported that cochlear damage may be then responsible of tinnitus possibly due to several mechanisms: (i) a "discordant damage" of Inner Hair Cells (IHC) and Outer Hair cells (OHC) in other words a prevalent damage of OHCs that only later, or only partially affects IHCs; (ii) the alteration of the coupling between the tectorial membrane and the OHCs; (iii) a dysregulation in the release of neurotransmitter from damaged IHCs [4].

In all the above conditions, hair cell injury and loss can also be followed by a retraction of the peripheral processes of the auditory nerve and, later, by a gradual degeneration of spiral ganglion neurons, and this can create further damage to the auditory pathways [18, 39].

Current evidence of the involvement of reactive oxygen species in the diseases of the inner ear (and auditory pathways) may therefore represent a possible therapeutic approach (by the use of antioxidants) to tinnitus secondary to hearing loss (i.e., due to acoustic trauma) [3, 4]. In particular, possible future pharmacological interventions could be aimed to: (i) interrupt the process of lipid peroxidation, thereby preserving the integrity of cell membranes (i.e., lazaroids), (ii) prevent noise induced damage mediated by ischemia/reperfusion (i.e., pentoxifylline and sartran), and (iii) to arrest the mechanisms of cellular apoptosis (i.e., using JNK inhibitors) [3]. These pharmacological interventions could therefore inhibit the onset/progression of hearing loss mediated by ROS, as well as reduce the perception of tinnitus [3].

Oxidative Stress and Idiopathic Tinnitus: Possible Hypothesis

The pathophysiological mechanisms related to the genesis and the persistence of idiopathic tinnitus is still controversial and therefore a rational pharmacological approach to this disease cannot be proposed yet. Moreover, there are still scarce studies in the literature among the aetiology of idiopathic tinnitus.

Some have hypothized that idiopathic tinnitus could result from an aberrant neural activity in any location along the auditory pathway (from the cochlear nerve to the dorsal cochlear nucleus to the auditory cortex) and that its persistence could be related to the involvement of limbic and emotional network [40]. The genesis of idiopathic tinnitus could then involve numerous etiological factors such as imbalance between excitatory and inhibitory neurotransmitters in the midbrain, hyperactivity of the auditory cortex and the dorsal cochlear nucleus, abnormal ciliar' hair cells activity, irritation of the cochlear or vestibular nucleus, and hyper-excitability of spiral ganglion neurons, which could also be induced by oxidative stress damage [2]. Dysfunction from the central cortex to the inner ear apparatus is increasingly thought to be related to biochemical pathway abnormalities and to free radical–induced oxidative damage [2]. Also, there is some evidence that psychological stress itself may cause oxidative damage in vivo, both in animal models and in humans; this phenomenon could be linked to biochemical pathway abnormalities in the midbrain and cortex and could then be linked to tinnitus [36, 40, 41, 42, 43].

So, there are some reports in the literature that attempt to assign to oxidative stress a role in the pathophysiology of idiopathic tinnitus. Particularly, Savastano et al. (2007) have observed elevated levels of ROS in the cerebral venous blood (internal jugular vein) of patients suffering by idiopathic tinnitus; a therapy with antioxidant agents administered orally (beta-carotene, vitamin C and E) resulted, in their study group of 31 patients, in a significant reduction of circulating ROS level, just after 48 hours from the beginning of the therapy, as well as in a reduction of the subjective tinnitus intensity (no change in auditory threshold has been observed) [44].

According to other studies, idiopathic tinnitus may be due to endothelium dysfunction within the cochlear microcirculation. They observed high levels of oxidative stress markers (as malondialdehyde, 4-hydroxynonenal, glutathione peroxidase, nitric oxide, L-ornithine, thrombomodulin and von Willebrand factor) in the cerebral venous blood of patients with idiopathic tinnitus; they concluded that oxidative stress may then be responsible for endothelial damage within the cerebro-vascular district and therefore within the inner ear microcirculation [45].

The aetiopathogenesis of idiopathic tinnitus is still controversial. More studies are needed in order to improve our knowledge about the pathophysiology of this condition in near future, and therefore to develop a causal therapeutic approach.

Conclusion

Oxidative stress has been implicated in the pathogenesis of some inner ear diseases, and therefore several attempts have been made to protect cochlear cells from reactive oxygen species with antioxidants. Nevertheless, as only few experimental data are available, more studies and randomized controlled trials are necessary in order to determine which could be the advantages of the antioxidant therapy in humans.

Concerning idiopathic tinnitus, it is still necessary to understand the etiopatogenetic mechanism underlying this disease, in order to then develop a causal approach.

References

[1] Holley MC. Hair cells regrowth. *International congress series 2003.* 1254: 1-6

[2] Haase GM, Prasad KN, Cole WC, Baggett-Strehlau JM, Wyatt SE. Antioxidant micronutrient impact on hearing disorders: concept, rationale, and evidence. *Am J Otolaryngol.* 2011 Jan-Feb;32(1):55-61.

[3] Ciorba A, Astolfi L, Martini A. Otoprotection and inner ear regeneration. *Aud Med,* 2008; 6: 170-175

[4] Ciorba A., Gasparini P., Chicca M., Pinamonti S., Martini A. Reactive oxygen species in human inner ear perilymph *Acta Otolaryngol.* 2010;130(2):240-6.

[5] Rybak LP, Whitworth CA. Ototoxicity: therapeutic opportunities. Drug Discovery Today 2005;10:1313-1321.

[6] Jiang H, Sha SH, Forge A, Schacht J. Caspase-independent pathways of hair cell death induced by kanamicin in vivo. *Cell death and Differentiation* 2006;13:20-30.

[7] Cheng AG, Cunningham LL, Rubel EW. Mechanisms of hair cell death and protection. *Current opinion in otolaryngology & Head and Neck Surgery* 2005;13:343-8.

[8] Cunningham LL, Cheng AG, Rubel EW. Caspase activation in hair cells of the mouse utricle exposed to neomycin. *J Neurosci.* 2002 Oct 1;22(19):8532-40.

[9] Budihardjo I, Oliver H, Lutter M, Luo X, Wang X (1999) Biochemical pathways of caspase activation during apoptosis. *Annu Rev Cell Dev Biol* 15:269–290.

[10] Wang W, Grimmer JF, van de Water T, Lufkin T. Hmx2 and Hmx3 homeobox genes direct development of the murine inner ear and hypothalamus and can be functionally replaced by drosophila hmx.. *Developmental Cell* 2004;7:439–453.

[11] Jiang H, Sha SH, Forge A, Schacht J. Caspase-independent pathways of hair cell death induced by kanamycin in vivo. *Cell Death Differ.* 2006 Jan;13(1):20-30.

[12] Nakagawa T, Yamane H, Takayama M, Sunami K, Nakai Y. Apoptosis of guinea pig cochlear hair cells following chronic aminoglycoside treatment. *Eur Arch Otorhinolaryngol.* 1998;255(3):127-31.

[13] 13 Pan, J.S.; Hong, M.Z.; Ren, J.L. Reactive oxygen species: A double-edged sword in oncogenesis. *World J. Gastroenterol.*, 2009, 15, 1702-1707.

[14] Baguley DM. Mechanisms of tinnitus. *British Medical Bullettin* 2002: 63: 195-212.

[15] Ruan, R.S. Possible roles of nitric oxide in the physiology and pathophysiology of the mammalian cochlea. *Ann. N.Y. Acad. Sci.*, 2002, 962, 260-274

[16] Sha, S.H.; Chen, F.Q.; Schacht, J. Activation of cell death pathways in the inner ear of the aging cba/j mouse. *Hear. Res.*, 2009, 254, 92-99.

[17] Sha, S.H.; Taylor, R.; Forge, A.; Schacht, J. Differential vulnerability of basal and apical hair cells is based on intrinsic susceptibility to free radicals. *Hear. Res.*, 2001, 155, 1-8.
[18] Poirrier AL, Pincemail J, Van Den Ackerveken P, Lefebvre PP, Malgrange B. Oxidative stress in the cochlea: an update. *Curr Med Chem*. 2010;17(30):3591-604.
[19] Priuska, E.M.; Schacht, J. Formation of free radicals by gentamicin and iron and evidence for an iron/gentamicin complex. *Biochem. Pharmacol.*, 1995, 50, 1749-1752.
[20] Sha, S.H.; Schacht, J. Formation of reactive oxygen species following bioactivation of gentamicin. *Free Radic. Biol. Med.*, 1999, 26, 341-347.
[21] Lesniak, W.; Pecoraro, V.L.; Schacht, J. Ternary complexes of gentamicin with iron and lipid catalyze formation of reactive oxygen species. *Chem. Res. Toxicol.*, 2005, 18, 357-364.
[22] Jiang, H.; Sha, S.H.; Forge, A.; Schacht, J. Caspase-independent pathways of hair cell death induced by kanamycin in vivo. *Cell Death Differ.*, 2006, 13, 20-30.
[23] Ylikoski, J.; Xing-Qun, L.; Virkkala, J.; Pirvola, U. Blockade of cjun n-terminal kinase pathway attenuates gentamicin-induced cochlear and vestibular hair cell death. *Hear. Res.*, 2002, 163, 71-81.
[24] Wang, J.; Van De Water, T.R.; Bonny, C.; de, R.F.; Puel, J.L.; Zine, A. A peptide inhibitor of c-jun n-terminal kinase protects against both aminoglycoside and acoustic trauma-induced auditory hair cell death and hearing loss. *J. Neurosci.*, 2003, 23, 8596-8607.
[25] Rybak, L.P.; Somani, S. Ototoxicity. Amelioration by protective agents. *Ann. N.Y. Acad. Sci.*, 1999, 884:143-51., 143-151.
[26] Rybak, L.P.; Whitworth, C.A.; Mukherjea, D.; Ramkumar, V. Mechanisms of cisplatin-induced ototoxicity and prevention. *Hear. Res.*, 2007, 226, 157-167.
[27] Yamane, H.; Nakai, Y.; Takayama, M.; Iguchi, H.; Nakagawa, T.; Kojima, A. Appearance of free radicals in the guinea pig inner ear after noise-induced acoustic trauma. *Eur. Arch. Otorhinolaryngol.*, 1995, 252, 504-508.
[28] Yamane, H.; Nakai, Y.; Takayama, M.; Konishi, K.; Iguchi, H.; Nakagawa, T.; Shibata, S.; Kato, A.; Sunami, K.; Kawakatsu, C. The emergence of free radicals after acoustic trauma and strial blood flow. *Acta Otolaryngol. Suppl.*, 1995, 519, 87-92.

[29] Yamashita, D.; Jiang, H.Y.; Schacht, J.; Miller, J.M. Delayed production of free radicals following noise exposure. *Brain Res.*, 2004, 1019, 201-209.

[30] Miller, J.M.; Brown, J.N.; Schacht, J. 8-iso-prostaglandin f(2alpha), a product of noise exposure, reduces inner ear blood flow. *Audiol. Neurootol.,* 2003, 8, 207-221.

[31] Ohinata, Y.; Miller, J.M.; Altschuler, R.A.; Schacht, J. Intense noise induces formation of vasoactive lipid peroxidation products in the cochlea. *Brain Res.*, 2000, 878, 163-173.

[32] Staecker, H.; Zheng, Q.Y.; Van De Water, T.R. Oxidative stress in aging in the c57b16/j mouse cochlea. *Acta Otolaryngol.*, 2001, 121, 666-672.

[33] Jiang, H.; Talaska, A.E.; Schacht, J.; Sha, S.H. Oxidative imbalance in the aging inner ear. *Neurobiol. Aging,* 2007, 28, 1605-1612.

[34] McFadden, S.L.; Ding, D.; Salvi, R. Anatomical, metabolic and genetic aspects of age-related hearing loss in mice. *Audiology*, 2001, 40, 313-321.

[35] Capaccio P, Pignataro L, Gaini LM, Sigismund PE, Novembrino C, De Giuseppe R, Uva V, Tripodi A, Bamonti F. Unbalanced oxidative status in idiopathic sudden sensorineural hearing loss. *Eur Arch Otorhinolaryngol.* 2012 Feb;269(2):449-53.

[36] Mazzoli M. *Complimentary tinnitus therapies; Texbook of Tinnitus:* Moller AR, Languth B, DeRidder D, Keinjung T Editors, Springer Ch 92, pp 733-747.

[37] Calabrese V, Cornelius C, Maiolino L, Luca M, Chiaramonte R, Toscano MA, Serra A. Oxidative stress, redox homeostasis and cellular stress response in Ménière's disease: role of vitagenes. *Neurochem Res.* 2010 Dec;35(12):2208-17.

[38] Bovo R, Ciorba A, Martini A. The diagnosis of autoimmune inner ear disease: evidence and critical pitfalls. *Eur Arch Otorhinolaryngol.* 2009 Jan;266(1):37-40.

[39] Webster, M.; Webster, D.B. Spiral ganglion neuron loss following organ of corti loss: A quantitative study. *Brain Res.*, 1981, 212, 17-30.

[40] Jastreboff PJ, Hazell JW. A neurophysiological approach to tinnitus: clinical implications. *Br J Audiol.* 1993;27(1):7–17.

[41] Liu, J, Wang, X, Shigenaga, MK, Yeo, HC, Mori, A and Ames, BN, Immobilization sffess causes oxidative damage to lipid, protein and DNA in the brain of rats *FASEB J*, 1996 10:1532-1538.

[42] Sivonova, M, Zitnanova, I, Hlincikova, L, Skodacek, I, Trebaticka, J and Durackova, Z, Oxidative sftess in university students during examinations *Stress*, 2004 7:1 83-1 88.

[43] Mercanoglu, G, Safran, N, Uzun, H and Eroglu, L, Chronic emotional stress exposure ilcreases infarct size in rats: the role of oxidative and nitrosative damage in response to sympathetic hyperactivity *Methods Find Exp Clin Pharmacol*, 2008 30 (10) :7 45 --7 52.

[44] Savastano M, Brescia G, Marioni G. Antioxidant therapy in idiopathic tinnitus: preliminary outcomes. *Arch Med Res*. 2007;38(4):456-9.

[45] Neri S, Signorelli S, Pulvirenti D, Mauceri B, Cilio D, Bordonaro F, Abate G, Interlandi D, Misseri M, Ignaccolo L, Savastano M, Azzolina R, Grillo C, Messina A, Serra A, Tsami A. Oxidative stress, nitric oxide, endothelial dysfunction and tinnitus. *Free Radic Res*. 2006; 40(6):615-618.

In: Tinnitus
Editors: F. Signorelli and F. Turjman
ISBN: 978-1-63117-556-5

Chapter 2

Tools for Tinnitus Measurement: Development and Validity of Questionnaires to Assess Handicap and Treatment Effects

Kathryn Fackrell[1], Deborah A. Hall[1], Johanna Barry[2] and Derek J. Hoare[1]
[1]NIHR Nottingham Hearing Biomedical Research Unit, Nottingham, UK
[2]MRC Institute of Hearing Research, Nottingham, UK

Abstract

Tinnitus is a chronic condition that affects about 15% of the population and up to one in three older adults. For some it is a mild annoyance, for others it can be extremely distressing and can significantly deteriorate quality of life. Perceptual characteristics are a poor indicator of clinical need. Clinicians and researchers alike rely on self-report or questionnaires to quantify the severity of an individual's tinnitus and to gauge the changes in tinnitus severity or tinnitus-related handicap over time or after clinical intervention. This book chapter evaluates the psychometric properties of five tinnitus questionnaires; Tinnitus Handicap Questionnaire, Tinnitus Reaction Questionnaire, Tinnitus Questionnaire, Tinnitus Handicap Inventory, and Tinnitus Functional

Index. We critically appraise the development process, validation, and responsiveness. We consider the true utility of each questionnaire to measure the short and long term consequences of tinnitus.

Introduction

Evidence-based assessment and treatment of tinnitus is important (e.g., Department of Health, 2009). However, tinnitus is notoriously difficult to measure objectively because it is an experiential phenomenon and can significantly differ between individuals. Objective measures include matching the pitch and loudness of tinnitus to an external sound. However, there is only a weak relationship between the psychoacoustic properties of sound (i.e., pitch or loudness) and its functional impact on the individual. Across those individuals with similar psychoacoustic attributes, there can be significant variability in self-reported handicap or the domains of impairment that are attributed to their tinnitus. It is not possible to determine or predict tinnitus distress based on the tinnitus pitch and loudness (Zeman et al., 2011; Andersson et al., 2005; Jakes et al., 1985). Perceptual characteristics therefore are not a good indicator of clinical need.

Tinnitus is multidimensional. Different symptom domains include emotional distress, problems with sleep, concentration or quality of life (QoL). Comorbidities in psychological well-being, such as stress, generalised anxiety and depression, and cognitive impairments profoundly impact on daily functioning (Stevens et al., 2007; Newman et al., 2011; Robinson et al., 2003). The degree to which tinnitus distress is perceived by patients can depend on the impact that tinnitus has on these factors or vice versa. For some it is a mild annoyance, for others it can be extremely distressing and significantly deteriorates quality of life.

Self-report measures, especially questionnaires, are the primary way to quantify the severity of an individual's tinnitus and to assess the changes in tinnitus severity or tinnitus-related handicap over time or after clinical intervention (Meikle et al., 2007). Although, it is not compulsory to include a standard questionnaire in clinical practice (Hesser, 2000), most clinicians in NHS audiology departments in England do use questionnaires to assess tinnitus severity (67% out of 138 respondents; Hoare et al., 2012). The Department of Health (DH) guidelines recommend the use of validated tinnitus questionnaires; namely the Tinnitus Handicap Inventory (THI) (Newman et al., 1996) and Tinnitus Questionnaire (TQ) (Hallam, 1996; 2008) for identifying tinnitus severity at intake assessment (Department of Health, 2009). Validated questionnaires can offer clinicians and researchers alike a systematic approach to quantifying tinnitus severity by distinguishing individuals who are bothered by their tinnitus from those who are not. This is a vital prerequisite for triaging patients effectively into the most appropriate interventions such as intensive management or education and advice. Questionnaires are also used for standardising selection criterion in research. For example, they identify people with similar degrees of tinnitus severity allowing the research to focus on a specific subtype.

Question items addressing different symptom domains can be categorised into subscales. Through analysing the responses in each subscale, clinicians can identify where the tinnitus distress lies and the specific domains of concern, so targeting the interventions. Tinnitus questionnaires can also be used as pre- and post-treatment outcome measures, therefore providing evidence of the changes that can occur in tinnitus severity. This is extremely important since there has been a move towards evidence-based commissioning within NHS services where healthcare professionals need to demonstrate to third-party payers the efficacy of management (NHS White Paper, 2010). Therefore there is a need to focus on evidence-based cost effective outcome measures that will improve patient experiences and the efficacy of interventions (NHS White Paper, 2010).

Evaluating and Validating a Questionnaire

Quality criterion for developing well validated questionnaires addresses four broad topics: reliability, validity, responsiveness and interpretation (Terwee et al., 2007). These are briefly summarised below.

Reliability

1. Internal consistency. The tinnitus questionnaire may assess different dimensions (subscales) and all the items in a subscale measure the same construct of tinnitus handicap.
2. Reproducibility – reliability. People with tinnitus can be distinguished from each other, despite measurement errors (relative measurement error)
3. Reproducibility – agreement. The tinnitus questionnaire score for individuals tested on multiple occasions over a short time period (1-2 weeks) are close to each other (absolute measurement error).

Validity

4. Content validity. The items in the tinnitus questionnaire are a correct and comprehensive reflection of the tinnitus handicap that the questionnaire is intended to measure.
5. Structural validity. The scores of the tinnitus questionnaire adequately reflect the construct of tinnitus handicap (i.e., the subscales explain at least 50% of the variance).
6. Construct validity. The scores demonstrate expected correlations between similar measures and expected differences between unrelated measures and these hypotheses should be defined a priori.

Responsiveness

7. Responsiveness. The tinnitus questionnaire is able to detect clinically important change over time.
8. Floor and ceiling effects. Few respondents achieve the lowest or highest possible score so that the tinnitus questionnaire does not compromise responsiveness, nor content validity and reliability.

Interpretation

9. Interpretability. It is possible to assign qualitative meanings to the quantitative scores, preferably with an indication of what change represents a minimal clinically important change for the patient group.

The remainder of this section considers and describes the techniques underlying each of the main quality criteria. It is recommended that the techniques are conducted from development to validation in the order presented below (Bland and Altman, 2002; Streiner & Norman, 2008). Table 1 provides a summary and definition of the different evaluation and validation techniques introduced in this review, and is intended to be used as a reference.

Step One: Ensuring the Questionnaire Content

As a useful general guide, developers should assess content validity to ensure that they (1) properly define what is being measured, (2) confirm that all aspects of the questionnaire (item choice, response options, and structure) are reviewed and judged by a panel of experts, (3) clearly report the development process, (4) ensure that any refinements to questionnaires undergo further evaluations, and (5) empirically validate the factorial structure (Fitzpatrick, 1983; Haynes et al., 1995; Streiner and Norman, 2008).

Table 1. Summary and definition of the evaluation and validation techniques described in this review

Validation technique	Definition
Content validity	Content validity is used during the initial questionnaire development to assess the degree in which all facets, such as items, response formats, and instructions are relevant to and represent the target construct (concept, attribute or variable) of the questionnaire. It is based on the qualitative judgements of experts, preferably using a 5 -7 point evaluation scales for agreement.
Frequency response distributions	Frequency response distribution provides information on the items ability to discriminate between individuals and to show sensitivity to change. It is the frequency count of a specific rating assigned to each item.
Cronbach's alpha	Cronbach's alpha provides a measure of the internal consistency. Inter-item correlations are an estimate of the proportion of variability due to the "true score" rather than measurement error. It is assumed that all of the items are tapping into the same construct and therefore positively correlate with each other. It is measured by essentially splitting the data in two in every possible way, and calculating the correlation coefficient for each split, therefore providing an average of these two values.
Inter-item correlations	Inter-item correlations provide a measure of internal consistency, assesses the association between pairs of items. This is an important component of item analysis as it highlights the inter-relatedness of the items and provides an indication of redundant items that are highly correlated. It is measured by correlating each item on the questionnaire with every other item.
Item-total correlations	Item-total correlations provide a measure of internal consistency, assessing the association between an item and the total score. It checks that the item is measuring the same construct as the test. It is measured by calculating the correlation coefficients of individual items with the questionnaire total, whilst omitting that item. It is computed using all the remaining items on the test.
PCA: Orthogonal varimax rotations	Orthogonal rotations improve factor extraction, maximising item loading. It is used to rotate the factors when the underlying factors are assumed to be independent of each other.

Table 1. (Continued)

Validation technique	Definition
PCA: Oblique rotations	Oblique rotations improve factor extraction, maximising item loading. This is a linear rotation that assumes that underlying factors are related or correlated with each other.
Construct validity	Questionnaires designed to tap into health behaviour constructs should be based on underlying theories and therefore should measure aspects that are consistent with this theory and construct. Construct validity assesses the extent to which a questionnaire actually measures the target construct it purports to measure. This type of validity consists of two components; convergent and discriminant validity.
Convergent validity	A component of construct validity. Convergent validity measures the extent to which the construct of a new questionnaire corresponds with other questionnaire constructs that are theoretically similar. It is measured by calculating the correlations coefficients between the questionnaires, and assessing the strength of the association.
Discriminant validity	A measure of construct validity. Assesses the extent to which the underlying construct of the new questionnaire can differentiate between constructs that are theoretically independent. It is measured by calculating the correlations coefficients between the questionnaires, and assessing the strength of the association.
Test-retest reliability	Assesses the consistency and stability of the questionnaire over time. Test-retest reliability determines the extent to which the scores maintained standing from test to retest. It is measured by calculating the correlations coefficients of two set scores either for a single item, subscale or total scale from the two different administrations points in time.
Minimal clinically important change	The clinical significance of a treatment requires knowledge about the minimal change in total scores of the specific questionnaire used. The minimal clinically important change score refers to the smallest change in total scores between baseline and final assessment that can be considered clinically relevant and be attributed to treatment benefits. This score is considered important for the design of clinical trials, the interpretation of the scores for the clinical significance of a treatment and its ability to meet standards of efficacy.
Global question	Usually a single question that is used to measure one aspect of the health behaviour, such as symptom severity, using a 5-point or 7-point response scale. Global questions are anchor-based techniques that provide a way to assess the corresponding changes in questionnaire total scores.

The order of the information corresponds to each step in evaluating and validation section. PCA: Principal Component Analysis.

Step Two: Validating the Questionnaire Structure

Validation provides a theoretical basis for the questionnaire construct (i.e., the concept, attribute or variable that is the target of measurement), Principal Component Analysis (PCA) is a popular technique for identifying the underlying domains of a questionnaire (Field, 2009). Essentially PCA identifies sets of correlated variables (Factors) within the questionnaire items. The factorial structure is informed by the inter-item relationships, computed from the inter-item correlations (factor loadings) between each variable in the analysis (Floyd, 1995). Factor interpretations are improved by rotating the data. These rotations can either be oblique or orthogonal in form. They are designed to maximise item loading on the extracted factor whilst reducing the loading on the other factors. A 'good' tinnitus questionnaire would be expected to cover all the important domains of tinnitus handicap (e.g., Kennedy et al., 2004). In other words, it would have a multifactor structure.

The internal structure ("internal consistency") is analysed using Cronbach's alpha, inter-item and item-total correlations, to assess the consistency of the item content so ensuring that all the items in a questionnaire are measuring the same underlying construct (Clark and Watson, 1995; Cortina, 1993; Table 1). For instance, a Cronbach's alpha score $\alpha > 0.7$ indicates acceptable internal consistency (Peterson, 1994; Table 2). In other words, the items all contribute to the overall construct.

Table 2. Category definitions of statistical tests: Cronbach's alpha and correlation coefficient

Type of test	Score	Category definition
Cronbach's alpha (α)	0.90 – 1.00	Extremely high
	0.80 – 0.89	Good
	0.70 - 0.79	Acceptable
	0.60 – 0.69	Questionable
	0.50 – 0.59	Poor
	< 0.50	Unacceptable
Correlation coefficient score (r)	0.80 – 1.00	Extremely strong
	0.60 – 0.79	Strong
	0.30 – 0.59	Moderate
	0.00 – 0.29	Weak

Step Three: Validating the Questionnaire Construct

To ascertain relevant and useful interpretations that 'truly' reflect the individual's perceived ability, the questionnaire should reliably measure the target construct health behaviour (Newman and Sandridge, 2004). A questionnaire should be assessed for the extent to which it actually measures the target construct it purports to measure ("construct validity"). Any new questionnaire should be compared, using correlation coefficients, to existing questionnaires that measuring constructs that are theoretically similar ("convergent validity") or constructs that are independent ("discriminant validity"). For example, a new tinnitus questionnaire would be expected to show relatively high correlations (i.e., high convergence) with other tinnitus questionnaires, and relatively low correlations (i.e., high discriminance) with generalised depression, anxiety and hearing handicap questionnaires because these represent theoretically distinct constructs (see Table 2).

Step Four: Validating the Interpretations of the Scores

It is important to be able to reliably grade tinnitus severity if treatment efficacy or treatment needs are to be established. It is also beneficial to interpret questionnaire scores according to categories that provide a clinical meaning to the numerical score. This enables clinicians and researchers alike to identify and quantify tinnitus severity. There does not appear to be a universally accepted guideline on how best to define categories and develop a grading system. One technique is to conduct quartile analysis on normative datasets. The quartile analysis simply divides the patient population into four categories based on the distribution of the total score before treatment. However, quartile scores are limited. Although often based on large normative datasets, important information related to patient experience is omitted. This could be provided through using a tinnitus severity global question to compare the perceived tinnitus severity to the overall scores (i.e., a reference anchor).

Step Five: Validating the Questionnaires Sensitivity to Change (Responsiveness)

Validating sensitivity should initially begin during questionnaire development. Item selection should include items that are sensitive to changes and the response scale should also be sensitive to small changes. Since the amount of response options are proportional to the items sensitivity to change, high resolution numerical scales (0-10) are recommended (Kirshner and Guyatt, 1995; Meikle et al., 2007). Changes in observed questionnaire scores should reliably and sensitively reflect change in the health behaviour being measured over time or after intervention.

To measure the consistency and stability of the questionnaire scores over time, test-retest reliability is used to assess the extent to which the scores are maintained from test to retest (without any interventions between the tests) (DeVon et al., 2007). Clinically meaningful treatment effects can impact upon the management of patients, and the ability to meet the requirements set by patients (Meikle et al., 2007). In order to measure the sensitivity of the questionnaire scores to changes in health behaviour, the variation that is clinically meaningful has to be separated from the measurement error that you naturally see between test and retest. To address this, the "minimal clinically important change" concept was developed (Jacobson et al., 1986; Jacobson et al., 1999; DeVet et al., 2006). This is defined as the value at which the change becomes clinically relevant to the patient. Two different approaches can be used to provide a comparison for the mean before-after treatment scores. (1) Distribution-based approaches are based on the statistical properties of the sample, i.e., the effect size (the difference in mean change and standard deviation change) (DeVet et al., 2006; Revicki et al., 2008). (2) Anchor-based approaches use an external indicator, such as global question response categories, to provide a reference of change (DeVet et al., 2006; Revicki et al., 2008).

Tinnitus Questionnaires

Over the past 30 years, a range of questionnaires have been developed to scale the severity and impact of tinnitus. These identify a number of dimensions associated with the complex construct of tinnitus severity, relating to sensory, behavioural and emotional reactions.

This section provides an in-depth evaluation of the psychometric properties of five tinnitus questionnaires, critically appraising their process of development, validation, and responsiveness. It focuses on the Tinnitus Handicap Questionnaire (THQ; Kuk et al., 1990), Tinnitus Questionnaire (TQ;

Hallam et al., 1988; Hiller and Goebel, 1992), Tinnitus Reaction Questionnaire (TRQ; Wilson et al., 1991), Tinnitus Handicap Inventory (THI; Newman et al., 1996) which appear in practice recommendations by the UK Department of Health and the international Tinnitus Research Initiative, and finally one of the most recent questionnaires to be developed; the Tinnitus Functional Index (TFI; Meikle et al., 2012).

In this section, the five questionnaires are presented in the order in which they were developed. The five questionnaires are evaluated following the validation steps above. Each questionnaire section starts with sections on initial questionnaire development (validation step one), whether the questionnaire covers all important tinnitus domains (validation step two), the two components of construct validity (validation step three), the ability of the questionnaire to grade tinnitus severity (validation step four) and the ability of the questionnaire to be responsive to treatment (validation step five).

Tinnitus Questionnaire (TQ)

From the best available information, it appears that the Tinnitus Questionnaire (TQ) was developed from the Tinnitus Effects Questionnaire (TEQ) (Hallam et al., 1988). However, the literature is unclear; there is no explicit description of the connection between these two questionnaires (see Henry and Wilson, 1998; Kennedy et al., 2004). Therefore, it is assumed that the TQ development started with Hallam et al. in 1988. Evidence from literature on the TEQ (i.e., Henry and Wilson, 1998) where the description of the questionnaire structure and scoring matches that of the TQ will be reviewed.

The TQ was developed by researchers at Royal National Throat Nose and Ear Hospital, London (Table 3; Hallam et al., 1988). Primarily designed to measure tinnitus severity, it is also used to evaluate change and to examine the relationship of different facets of complaint and other psychological variables to tinnitus (Hallam, 2008). For each item, individuals indicate the level of agreement using one of three response options; not true (0), partly true (1) and true (2). The total score is rescaled from the weighted sum of the items used in each subscale (41 items) so that the global score ranges from 0 –82, with higher scores indicating increased tinnitus distress (Table 4).

Table 3. Characteristics and psychometric properties of the five widely-used tinnitus questionnaires

Questionnaire (Author, year)	Response options	No of items Subscales	Reported aims of measure	Psychometric properties	No. of participants
Tinnitus Questionnaire (TQ) (Hallam et al., 1988; Hallam, 1996, 2008)	**3 levels:** True, partly true, not true	**52 items** **5 subscales:** Emotional and cognitive distress, Intrusiveness, Auditory Perceptual difficulties, Sleep disturbance, Somatic complaints	Measures psychological aspects of tinnitus complaints and distress	Cronbach's Alpha $\alpha = 0.91$	106
				Test-retest reliability: $r = 0.94$[b]	60
Tinnitus Handicap Questionnaire (THQ) (Kuk et al., 1990)	**100 levels:** 100 = strongly agree, 0 = strongly disagree	**27 items** **3 subscales:** Physical, emotional & social effects of tinnitus, Hearing and communication ability, Individual's perception of tinnitus	Measure of patients' perceived degree of handicap due to tinnitus	Cronbach's Alpha $\alpha = 0.94$	275
				Test-retest reliability: $r = 0.89$[a]	32
Tinnitus Reaction Questionnaire (TRQ) (Wilson et al., 1991)	**5 levels:** Not at all to almost all the time	**26 items** **No clear subscales**	Measure psychological distress associated with tinnitus	Cronbach's Alpha: $\alpha = 0.96$	156
				Test-retest reliability: $r = 0.88$	43
Tinnitus Handicap Inventory (THI) (Newman et al., 1996)	**3 levels:** Yes Sometimes No	**25 items** **3 subscales:** Functional, Emotional, Catastrophic	Measures the level of perceived tinnitus severity	Cronbach's Alpha $\alpha = 0.93$	66
				Test-retest reliability: $r = 0.92$[c]	29
Tinnitus Functional Index (TFI) (Meikle et al., 2012)	**11 levels:** 0 – 10 Descriptive anchors vary between items	**25 items** **8 subscales:** Intrusiveness, Sense of control, Sleep, Cognition, Hearing, Relaxation, Emotional distress, Quality of life	Developed to measure both tinnitus severity and treatment-related changes	Cronbach's Alpha $\alpha = 0.97$[d]	336
				Test-retest reliability: $r = 0.78$[d]	27

[a] data from Newman et al., 1995.
[b] data from Hiller et al., 1994.
[c] data from Newman et al., 1998.
[d] 25-item prototype 2 data.

Table 4. The progression of the Tinnitus Questionnaire development

Year Version (Author)	No. of Items (factors)	Subscales	No. of items in subscales	Cronbach's alpha(α)	Scoring
Prototype (Hallam et al., 1988)	40 (3/4)	Emotional distress	20 items	N/A	N/A Remaining items:10
		Auditory perceptual difficulties	6 items		
		Sleep disturbance	4 items		
Final TQ (Hallam et al., 1988)	51 (3)	Sleep disturbance	13 items: 2, 4, 18, 20, 12, 31, 34, 35, 36, 39, 41, 48, 50	N/A	N/A Remaining items:17
		Emotional distress	15 items: 3, 5, 7, 10, 11, 13, 15, 19, 24, 27, 37, 39, 47, 43, 45		
		Auditory perceptual difficulties	6 items: 9, 14, 18, 26, 33, 38		
German version GHTQ (Hiller & Goebel, 1992)	52 (5/6)	Emotional distress*	12 items: 1, 5, 8, 11, 16, 18, 19, 20, 28, 37, 39, 41	α = .85	Scores range: 0-84 Global score based on 42 items (α = .93) Remaining items:12
		Cognitive distress*	8 items: 3, 13, 17, 21, 27, 43, 44, 47	α = .85	
		Intrusiveness	8 items: 5, 7, 10, 15, 20, 34, 35, 48	α = .75	
		Auditory perceptual difficulties	7 items: 2, 9, 14, 26, 33, 38, 50	α = .86	
		Sleep disturbance	4 items: 4, 12, 31, 36	α = .80	
		Somatic complaints	3 items: 22, 25, 51	α = .74	
Revised TQ (Hallam, 1996)	52 (5)	Emotional and cognitive distress	19 items: 3, 8, 13, 16, 17, 18, 19, 20, 21, 24, 27, 28, 30, 37, 39, 41, 43, 44, 47	α = .94	Scores range: 0-82 Overall score
		Intrusiveness	7 items:	α = .87	

Year Version (Author)	No. of Items (factors)	Subscales	No. of items in subscales	Cronbach's alpha(α)	Scoring
			5, 7, 10, 11, 15, 35, 45		based on 41 items (α = .95) Remaining items:11
		Auditory perceptual difficulties	7 items: 2, 9, 14, 26, 33, 38, 50	α = .89	
		Sleep disturbance	4 items: 4, 12, 31, 36	α = .81	
		Somatic complaints	4 items: 22, 25, 29, 34	α = .75	

TQ; Tinnitus Questionnaire, GHTQ Goebel and Hiller Tinnitus Questionnaire.
* Emotional and Cognitive distress can also be combined. Numbers that appear in bold are featured twice in the questionnaire, in different subscales.

Unfortunately, the British version - the TQ (Hallam) – has little published information available. Apart from Hallam et al. (1988), the only evidence available is in the 2008 TQ manual (Hallam, 2008). The 1996 TQ manual (Hallam), where the revisions of TQ internal structure are reported, is out of print. Any evidence provided of discriminant and convergent validity in 1988 is now redundant because of subsequent modifications to the questionnaire. For example, the revisions to the TQ in 1996 mean that "The present scoring of subscales [...] supersedes the earlier scoring system" (Hallam, 2008, p. 9). The revised TQ (Hallam 1996; 2008) presented an opportunity to conduct full psychometric validation. However, Hallam does not appear to have conducted any additional validation beyond that of validating the internal TQ structure. Evidence reported in the 2008 manual is based on unpublished data and redundant 1988 data (i.e., Hallam et al., 1988). In this review, we will refer to evaluation and validation evidence from the German translation (GHTQ; Hiller and Goebel, 1992; Goebel and Hiller, 1998). The global score ranges from 0 – 84 for the GHTQ, this is based on the 42 items in the subscales (out of a total 52 items) with two items are included twice (Q5 and Q20).

Initial Questionnaire Development

From the information provided by Hallam et al. (1988) on the development of the TQ (prototype/final versions), it is difficult to ascertain the process and quality of development. There is only a reference to the choice of items being based on clinical observations of the most commonly reported adverse effects and complaints associated with the multiple domains of tinnitus (see Hallam et al., 1988; Hallam, 2008).

The authors do refer back to the previous work by Jakes et al. (1985). They investigated the relationship between loudness and annoyance of tinnitus with a forced-choice questionnaire. Tinnitus handicap was found to have two dimensions (emotional distress and intrusiveness). It is not apparent from the literature how these two factors informed Hallam and colleauges' decisions on item choice and yet they are regularly referred to as the basis of the TQ (Hiller and Goebel, 1992; Henry and Wilson, 1998; Gerhards et al., 2004). No further information was reported on the item choice.

No empirical validation was conducted until 1996 in the UK (Hallam, 1996). This work produced a revised TQ (Table 4). Previous items were unchanged, but the questionnaire now included an additional item that was "accidently missed" out of the 1988 version due to clerical error (Hallam, 2008). To compute the global and subscale scores for the TQ only 41 out of the possible 52 items are used, and for the GHTQ only 42 items are used (Table 4). The remaining items are considered as baseline information that may be clinically useful (Hallam, 2008, p.7). In fact they could be considered a hindrance as they increase the time required to administer the questionnaire and there is limited evidence for their clinical relevance (Hiller and Goebel, 2004; Newman and Sandridge, 2004).

Questionnaire Structure

The TQ purports to encompass the major domains of tinnitus (Hallam et al., 1988; Hallam, 2008). Over the years the factorial structure of the TQ has been revised and consequently the number of domains purported to be covered by TQ has changed (Table 4) (Hiller and Goebel, 2004).

Factorial Structure

After conducting PCA, the 1988 version (Hallam et al. 1988) reported three factors based on 34 of the 51 items. These were i) emotional distress, (ii) auditory difficulties and (iii) sleep disturbance. There was no further empirical validation. There is no evidence for the internal consistency of each subscale and the selected 34 items of the TQ.

In 1992, Hiller and Goebel investigated the psychometric properties of the GHTQ (German translation). PCA with orthogonal varimax rotations, followed by factorial stability analysis, revealed six main factors for interpretation: (i) emotional distress, (ii) auditory difficulties, (iii) sleep disturbance, (iv) cognitive distress, (v) intrusiveness, and (vi) somatic complaints (Hiller and Goebel, 1992). The first three factors had previously been identified by Hallam et al. (1988). The intrusiveness factor had previously been identified by Jakes et al. (1985) but not by Hallam et al. (1988).

In 1996, Hallam reinvestigated the factorial structure of the TQ. Using the same techniques as Hiller and Goebel (1992), a six factor solution was revealed, but only five factors were considered reliable. These were (i) emotional and cognitive distress, (ii) auditory difficulties, (iii) sleep disturbance, (iv) intrusiveness, and (v) somatic complaints. Here, the emotional and cognitive factors previously reported by Hiller and Goebel (1992) were combined into a single factor. The 2008 manual provides a brief explanation of the 1996 revalidation, but at times the information about how the questionnaire items load onto each factor is contradictory and vague. For instance, different items are listed under each factor throughout the manual (Seydel et al., 2012).

Nevertheless, the five factor structure of the TQ has also been replicated in its Cantonese (Kam et al., 2009), Dutch and French (Meeus et al., 2007) translations. The factorial structure of the 52-items would suggest that the TQ (Hallam, 1996; 2008) measures five separate domains of tinnitus distress, but some questions remain about the reliability of this conclusion.

It is important to highlight that the PCA analyses described GHTQ and TQ have been based on all 52 TQ items, not on the 41 items that contribute to calculate the global score. This aspect of the statistical methodology could have a major impact on the resulting factorial structure of the global TQ score (Gerhards et al., 2004).

Indeed, following PCA on the GHTQ's 42 'overall score' items, Gerhards reported only one or two factors instead of the previously reported five-six factors (Gerhards et al., 2004). Furthermore, just under half (47%) of the items focus on the emotional problems, although the remaining four tinnitus domains appear to be more evenly distributed (Figure 1; Kennedy et al., 2004).

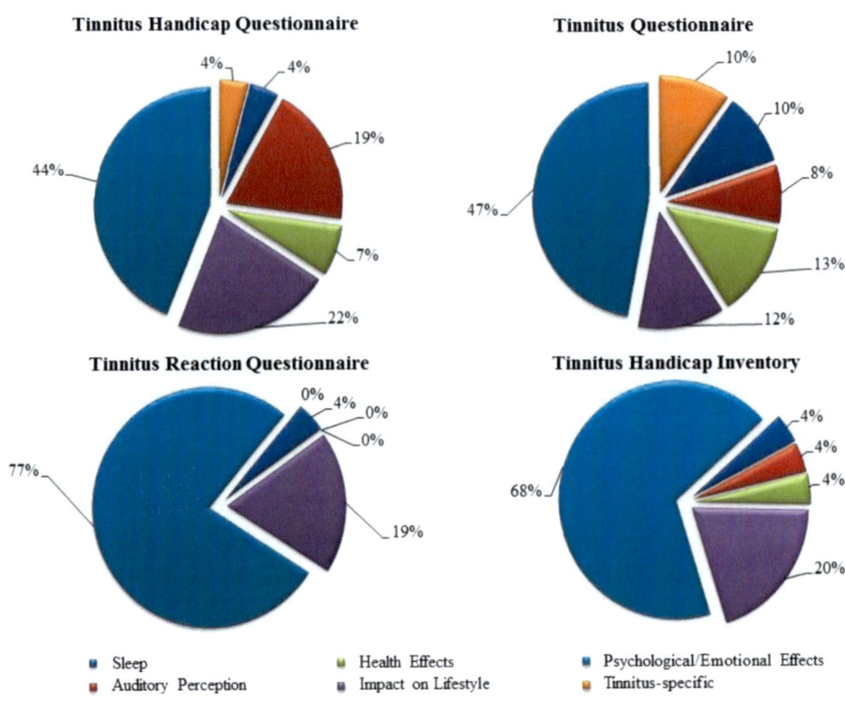

Figure 1. Questionnaires assessed on six domains with a different number of items focussed on each domain: The Tinnitus Handicap Questionnaire (THQ), Tinnitus Questionnaire (TQ), Tinnitus Reaction Questionnaire (TRQ), and Tinnitus Handicap Inventory (THI). Adapted from Kennedy et al. (2004).

Internal Consistency

The 2008 manual reported the internal consistency (Cronbach's alpha) of the TQ. Alpha estimates were extremely high for the total score and sleep disturbance, good for emotional distress, auditory difficulties and intrusiveness, and acceptable for somatic complaints (Table 4). These high scores would suggest there is little variance between the individual items (Cronbach, 1951).

Questionnaire Construct

Construct Validity: Convergent Validity

The TQ measures constructs that are comparable to other tinnitus-related questionnaires. High convergent validity for TQ and TEQ has been shown by the strong correlations with the THQ, (r = 0.75), the TRQ (r = 0.74) (Henry and Wilson, 1998; Robinson et al., 2003), and the THI total (r = 0.89) and subscale scores (r = 0.79 to r = 0.86) (Baguley et al., 2000; Robinson et al., 2003). The TQ also showed strong correlations with the Tinnitus Cognitions Questionnaire (TCQ) negative subscale (r = 0.65) (Wilson and Henry, 1998).

Construct Validity: Discriminant Validity

The TQ demonstrates moderate to high discriminant validity indicating that it does measure constructs that are distinct from more generalised symptoms. The TQ shows moderately correlations with the Beck's Depression Inventory (BDI; Beck et al., 1997) (r = 0.55) (Henry and Wilson, 1998; Robinson et al., 2003), the Hamilton Rating Scale for Depression (Hamilton, 1960) (r = 0.48), the Modified Somatic Perception Questionnaire scores (Main, 1983) (r = 0.46) and the Quality of Well-Being scale scores (Kaplan et al., 1996) (r = -0.37; Robinson et al., 2003). It has weak to moderate correlations with the Hopkins Symptom Checklist (SCL-90-R; Derogatis, 1977) (r = 0.26 to r = 0.39) (Hiller et al., 1994; Hiller and Goebel, 2004) and weak correlations with the Private Self-Consciousness Scale scores (Fenigstein et al., 1975) (r = 0.21; Robinson et al., 2003).

Interpretation of the Scores

Hallam (2008) suggests that quantifying tinnitus distress should be left to the clinicians' discretion. Nevertheless, a grading system has been developed for the GHTQ (Goebel and Hiller, 1998) which provides clinical meaning to the scores. Gerhards et al. (2004) conducted quartile analysis on the data collected from 683 tinnitus patients in Germany and confirmed the same range of scores in the four categories. These data are summarised in Table 5.

Table 5. Grading systems providing qualitative meanings to the quantitative scores

	Score	Tinnitus handicap	Description of symptom severity
THQ (Sullivan et al., 1993)	>22	Bothersome tinnitus	
GHTQ (Goebel & Hiller, 1998)	0 – 30	Mild	
	31 – 46	Moderate	
	47 – 59	Severe	
	60 – 84	Very severe (heavy burden)	
THI (McCombe et al., 2001)	0 – 16	Slight	"Only heard in a quiet environment, very easily masked. No interference with sleep or daily activities".
	18 – 36	Mild	"Easily masked by environmental sounds and easily forgotten with activities. May occasionally interfere with sleep but not daily activities".
	38 – 56	Moderate	"May be noticed, even in the presence of background or environmental noise, although daily activities may still be performed. Less noticeable when concentrating. Not infrequently interferes with sleep and quiet activities".
	58 – 76	Severe	"Almost always heard, rarely, if ever, masked. Leads to disturbed sleep pattern and can interfere with abilities to carry out normal daily activities. Quiet activities affected adversely. There should be documentary evidence of the complaint having been brought to the general medical practitioner. Hearing loss is likely to be present but its presence is not essential".
	78 – 100	Catastrophic	"All tinnitus symptoms at level of severe or worse. Should be documented evidence of medical consultation. Hearing loss is likely to be present but its presence is not essential. Associated psychological problems are likely to be found in hospital or general practitioner records".
TFI (Meikle et al., 2012)	< 25	Mild	
	25 – 50	Significant problems with tinnitus	
	> 50	Severe	

Sensitivity to Change

Hallam (2008) claimed that the "Chief application of the TQ is in the evaluation and auditing of psychological interventions for tinnitus" (p.7). The evidence for this claim appears somewhat limited because the TQ items were not selected to be specifically responsive to treatment-related changes. It consequently lacks the required resolution to detect small changes in scores (Meikle et al., 2008). In all versions of the TQ (Hallam et al., 1988; Hallam, 1996; 2008), no data has been provided to determine a clinically significant change score following intervention (Newman and Sandridge, 2004). Therefore, clinicians and researchers alike would be advised not to rely on the TQ for auditing purposes or to determine minimal significant change.

Goebel et al. (2006) have categorised treatment effects using the TQ. They concluded 6-14 point change in TQ scores was indicative of improvement, i.e., the patient was "responding" to the treatment. They also classified a change of ≥15 points as a "winning response", i.e., the patient was showing a large improvement. However, an alternative minimal clinically important change score for the GHTQ has been proposed by Adamchic et al. (2012). Using data from both the Randomized Evaluation of Sound Evoked Treatment of Tinnitus (RESET) trial study (N = 63) and Tinnitus Research Initiative (TRI) database (N = 694), Adamchic et al. (2012) analysed 757 patient responses on the GHTQ at baseline, clinical assessment and end of treatment. There was an average of 44 days between baseline and clinical assessment. Adamchic et al. (2012) used a combination of anchor-based and distribution-based techniques, such as Clinical Global Impression Improvement (CGI-I) rating, Receiver Operating Characteristic (ROC) curves and Standard Error Measurement (SEM) to produce the minimal clinically important change score for improvement. The CGI-I is an overall rating that uses up to seven response options to quantify treatment-related change in health behaviour, i.e., patients make a judgement on the total improvement of their tinnitus after treatment (Table 6a, 6b). For the analysis, the patient groups were formed according to the CGI scores. The seven CGI-I categories in the TRI database and the five CGI-I categories in the RESET database were combined into five groups; much better, minimally better, no change, minimally worse, much worse. The authors examined the predicted minimal clinically important change scores for each technique and choose the most representative score. The SEM scores estimated the lowest change score of -4.7 but the ROC curve predictions of a change score of -5 was considered most representative. This recommendation differs from that of Goebel et al. (2006). It's derivation from a larger dataset and using a combination of distribution-based and anchor-based methods, potentially offers a more precise estimate of meaningful change.

Table 6. Clinical Global Questions to determine patients judgement on perceived treatment-related change

a) TRI Clinical Global Impression	b) RESET Clinical Global Impression
Rate the total improvement of their tinnitus complaints compared to before the beginning of treatment.	**Verbal rating of their tinnitus loudness and annoyance for each ear where the tinnitus was perceived as compared to baseline.**
1. Very much better	1. Much better
2. Much better	2. Somewhat better
3. Minimally better	3. No change
4. No change	4. Somewhat worse
5. Minimally worse	5. Much worse
6. Much worse	
7. Very much worse	

c) TFI Global Question
All things considered, how is your overall tinnitus condition now, compared to your first visit to this clinic?
1. Much improved
2. Moderately improved
3. Slightly improved
4. No change
5. Slightly worse
6. Moderately worse
7. Much worse.

Summary

There is limited information available for the TQ, and the information that is available is somewhat poorly specified. The TQ appears to measure five separate domains of tinnitus distress. However, the validation of the factorial structure has generally not been conducted on the 41 items that contribute to the total TQ score and subscale scores. The TQ does have high construct validity and would appear to measure tinnitus severity, although patient experience was potentially overlooked during the development of the grading system. Recently there have been attempts to overcome the lack of evidence for minimal clinically important change scores, with some success.

Tinnitus Handicap Questionnaire (THQ)

The 27-item THQ was developed to comprehensively measure a patient's tinnitus handicap and to be sensitive to the changes in handicap over time (Kuk et al., 1990). It is claimed to measure three broad domains of tinnitus handicap: (i) the impact of tinnitus on social, emotional and physical aspects, (ii) hearing ability and unease, and (iii) the individuals' perception of their tinnitus (Table 3). For each item on the THQ, the individual is asked to indicate how much he/she disagrees or agrees with the statement with a number between 0 (strong disagree) and 100 (strongly agree). The total score is rescaled from the weighted sum of all the items in each subscale so that the global score ranges from 0 – 100, with a higher score indicating greater handicap.

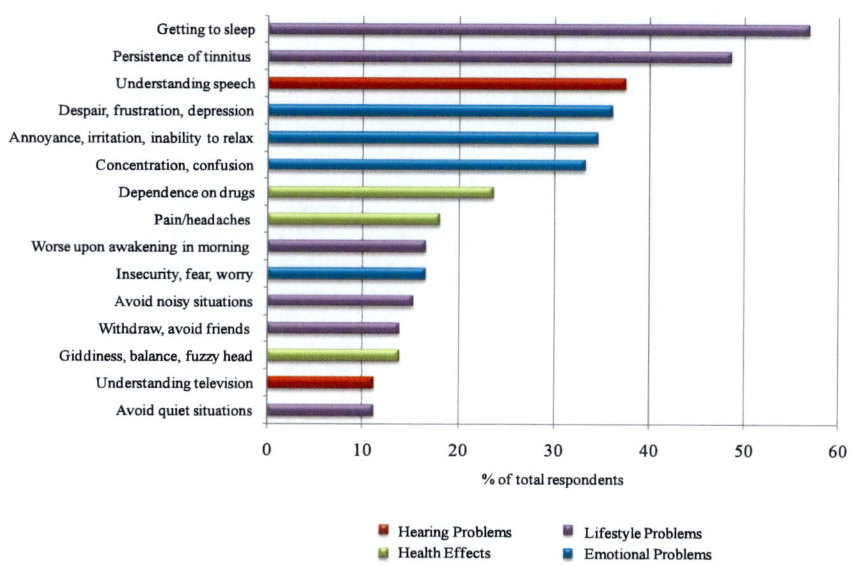

Figure 2. Fifteen most common difficulties reported by tinnitus patients in open-ended questionnaire. Adapted from Tyler and Baker (1983).

Initial Questionnaire Development

At the first step in the development, the authors obtained the 87 items directly from Tyler and Baker (1983; Figure 2) which were "arbitrarily" grouped into four domains that reflect tinnitus handicap (i.e., hearing, lifestyle, health, and emotion).

The authors briefly refer an additional domain on "others' reaction to tinnitus", but it is unclear how this additional domain was derived. Kuk et al. (1990) systematically reduced the initial 87 items to the final 27-item questionnaire, by sampling items that were sensitive to perceived tinnitus handicap. They did this by examining (i) response frequency distributions, (ii) inter-item correlations, and (iii) item-total correlations.

i. Items were eliminated based on their ability to differentiate between individual responses, i.e., the sensitivity of item. Items were assumed to be insensitive, if they were scored with either a 0 or 100 on the response options over 50% of the time. However, the authors do not clearly explain why they chose this criterion and it was applied inconsistently at this development stage. For example, the developers retained items with 50% "0" ratings because they "meet other criteria" (not specified).
ii. Highly correlated items were examined. The specific criterion value for high inter-item correlations was not reported, although an example of $r > 0.60$ was given as an example of redundant items. Multiple items that reflected the same situation, for example listening in quiet, were considered redundant. The most representative item of them was kept, whilst the rest of the redundant items were eliminated.
iii. Items were eliminated based on their "low" item-total correlations in order to maximise internal consistency. However, the specific cut-off value for a low item-total correlation was not reported. The authors considered the response frequency distributions (sensitivity) before eliminating the item based on the internal consistency scores. Two items were kept despite their low item-total scores.

Questionnaire Structure

Factorial Structure

Having conducted PCA and oblique rotations on the THQ, three factors were clearly identified (Kuk et al., 1990). Factor 1 represents social, emotional and physical functioning (15 items). Factor 2 represents hearing ability and unease (8 items), and factor 3 represents individual perception of tinnitus (4 items). Factors 1 and 2 were moderately correlated (r = 0.49), but both were only very weakly correlated (r = 0.10, r = 1.16 respectively) with factor 3. Items on factor 3 also resulted in extremely low item-total correlation scores (r = 0.15) suggesting that it is potentially measuring separate variables to factors 1 and 2. Finally, the majority of items loaded onto factor 1 with some overlap of items onto factor 2 indicating that the THQ is particularly sensitive to the social, emotional and physical functioning aspects of tinnitus distress. This is consistent with Kennedy et al.'s (2004) critical review. Kennedy designated the items according to the six categories highlighted by Tyler and Baker's open questionnaire study (1983) and found that the THQ places little emphasis on issues to do with sleep deprivation, cognitive disturbance or impact on health and well-being (Kennedy et al., 2004; Figure 1).

Internal Consistency

The THQ demonstrated extremely high Cronbach's alpha for both the THQ total ($\alpha = 0.94$) and for factor 1 ($\alpha = 0.94$). These high alpha values could reflect the low uniqueness of the items content (Cortina, 1993). Some items could essentially be asking the same question in a different guise and therefore reflecting unnecessary duplication of content (Streiner, 2003). This provides additional evidence of a factorial structure that mainly focuses on a limited number of tinnitus domains. Factor 2 had a good Cronbach's alpha ($\alpha = 0.88$) (Kuk et al., 1990). Factor 3 however, achieved unacceptable Cronbach's alpha ($\alpha = 0.47$). Kuk et al. (1990) attributed this unacceptable score to the small number of items in the subscale and the low inter-item correlation. Alternatively, factor 3 might measure unrelated items that are not reflecting the overall concept of "Individual perception of tinnitus". If this were true then it would not be contributing any additional value to the overall concept of tinnitus handicap (Cortina, 1993; Tavakol and Dennick, 2011).

Questionnaire Construct

Construct Validity: Convergent Validity

During the THQ development, the THQ displayed moderate correlations with average hearing thresholds (r = 0.52) and perceived tinnitus loudness (r = 0.57) (Kuk et al., 1990). Since then, the THQ has also been found to have strong correlations with the TQ (r = 0.75), the TRQ scores (r = 0.74) (Henry and Wilson, 1998) and the THI scores (r = 0.76; Robinson et al., 2003). Overall, the THQ demonstrates high convergent validity and therefore successfully measures aspects of tinnitus distress similar to other tinnitus questionnaires.

Construct Validity: Discriminant Validity

Low to moderate discriminant validity has been observed for the THQ. The THQ indicates moderate discriminant validity, displaying weak correlations with the Modified Somatic Perception Questionnaire scores (r = 0.37) the Private Self-Consciousness Scale (r = 0.11; Robinson et al., 2003), and moderate correlations with the life satisfaction scales (Lohmann, 1980) (r = 0.54), the physical health status subscale (r = 0.54), the general health subscale (Duke-UNC Health profile; Parkerson et al., 1981) (r = 0.54; Kuk et al., 1990) and the Quality of Well-Being scale (r = 0.48; Robinson et al., 2003).

Nevertheless contradictory evidence indicates rather low discriminant validity. For example, the THQ has been shown to highly correlate with the Zung Self-Rating depression scale (Zung, 1965) (r = 0.63; Kuk et al., 1990) and the BDI (r = 0.62; Robinson et al., 2003). Robinson et al. (2003) also reported that the THQ moderately correlated with the Hamilton Rating Scale for Depression (r = 0.57).

Overall, the evidence suggests that the THQ has low to moderate discriminant validity and that it may be susceptible to generalised emotional distress, rather than distress that is specific to the condition of tinnitus.

Interpretation of the Scores

It is important to be able to reliably grade tinnitus severity if treatment efficacy or treatment needs are to be established. No grading system has been developed for the THQ measurement scale. Although a score of over 22/100 points (or 600/2700 points in raw scores) has been suggested as a lower boundary for bothersome tinnitus (Sullivan et al., 1993), there does not appear to be reliable empirical evidence to support this claim (see Table 5). Kuk et al. (1990) suggested that one possible method of determining the severity of symptoms for individuals is to compare a THQ score to their published normative data from 275 tinnitus patients. For example, a mean score of 60 would inform a clinician that the individual's tinnitus was more severe than 80% of the tinnitus patients in this reference sample (Figure 3). Although this ranking is helpful in determining individual severity relative to others, it does not provide clinical interpretations of the scores (Newman and Sandridge, 2004; Kuk et al., 1990). For example, the mean score of 60 is not classified into a category, such as 'moderate tinnitus', that can inform the clinician of how severe the tinnitus is perceived and guide management.

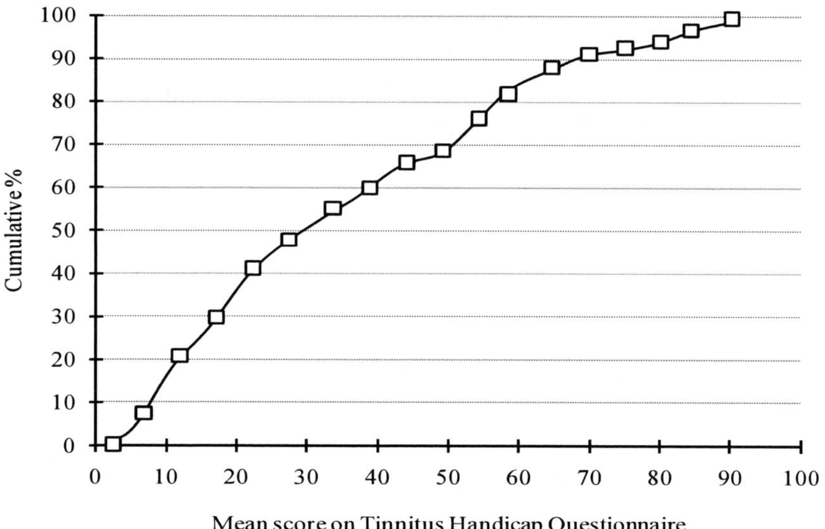

Figure 3. Cumulative distribution of total scores on the Tinnitus Handicap Questionnaire (THQ). Data obtained from 275 patient responses (Kuk et al., 1990). Adapted from Kuk et al. (1990).

Sensitivity to Change

The THQ was primarily developed to effectively evaluate treatment-related change, with less emphasis placed on its ability to discriminate between diagnostic categories (Kuk et al., 1990). The THQ has high resolution response options (0 -100), allowing for small changes in the scores to be detected. High test-retest reliability has been observed, with strong correlations between administrations for the total THQ scores (r = 0.89), factor 1 scores (r = 0.89) and factor 2 scores (r = 0.90) (Newman et al., 1995). Therefore, the global THQ score, and scores for factors 1 and 2 are stable across time. Changes can therefore probably be attributed to treatment-related effects. In contrast, factor 3 only showed moderate correlations (r = 0.50), suggesting that any changes shown over time could in fact be due to SEM rather than treatment-related effects.

Despite being specifically designed to maximise treatment responsiveness, Kuk et al. (1990) report no information on a minimal clinically important change score. The evidence for a minimal change score is dependent on the test-retest reliability data (Newman et al., 1995). From calculating the 95% confidence intervals (variance around the scores between administrations) with the SEM, Newman et al. (1995) conclude that a mean global THQ score would have to change by 21 points for the change to be considered clinically significant. However, this value maybe slightly compromised by lack of consistency and reproducibility shown by factor 3.

Summary

In conclusion, the development process of the THQ was informative and in general clearly reported, but there was a lack of clarity over the criterion scores for eliminating items. The THQ appears to cover some aspects of tinnitus distress but the emotional aspects seems to be the main contributor to global score. The usefulness and reliability of the three subscales is questionable. Factor 3 does not appear to add any further value to the questionnaire and is deemed to have a heterogeneous construct. Nevertheless, the THQ does show high convergent validity with other tinnitus questionnaires. There is no grading system to provide clinical meaning to the scores and a clinically meaningful change score has been suggested but it may be compromised by the limitations of factor 3.

Tinnitus Reaction Questionnaire (TRQ)

The 26-item Tinnitus Reaction Questionnaire was specifically designed to assess the psychological distress related to tinnitus, such as anger and depression (Wilson et al., 1991). Primarily developed to be an assessment of the effects of psychological interventions on tinnitus (treatment-related change), it is also claimed to distinguish between the levels of tinnitus-related distress. However, there is very little literature on the psychometric properties of the TRQ, with no apparent evidence for defining tinnitus distress values or a minimal clinically important change score. For each item, ratings are made on a 5-point scale with the anchors from: not at all (0) to almost all of the time (4). The total score ranges from 0 to 104, with a higher score denoting higher levels of distress (Table 3).

Initial Questionnaire Development

Critical evaluation of the development of the TRQ is difficult because of insufficient information provided in Wilson et al. (1991). For TRQ item choice, the authors briefly refer to the symptom categories reported by Tyler and Baker (1983; Figure 2) as the principal foundation, alongside the experience gained in previous work interviewing tinnitus patients on the efficacy of relaxation therapy (e.g., Ireland et al., 1985). However, it is not apparent exactly how this information was used to inform item development. It would appear that from the initial conception, the TRQ has only ever consisted of 26 items and so the number of items was never reduced like in the other cases. Although the authors did investigate response distributions, this was not used to inform the choice of items. No further information was provided on the item development and content.

Questionnaire Structure

Factorial Structure

The TRQ claims to measure aspects of psychological distress associated with tinnitus. It does not focus on the other domains of tinnitus distress discussed so far. The initial item-total correlations supported this, revealing high correlations between the items (ranged from $r = 0.44$ to 0.81) with the majority scoring $r > 0.70$, indicating that the TRQ measures a narrow range of tinnitus characteristics. PCA and orthogonal rotation revealed four factors related to psychological distress; (i) general distress (15 items), (ii) interference (9 items), (iii) severity (8 items), and (iv) avoidance (3 items) (Wilson et al., 1991). The analysis revealed that the majority of items loaded on more than one factor. For example, the majority of the items on the "interference" factor also load onto the "severity" factor. This indicates that the four factor solution is not representing distinctive subscales and that there is little value in looking at the factors individually.

Wilson et al. (1991) did also describe a two factor solution, where the majority of items loaded on factor 1, with six items loading on factor 2 (avoidance). The authors do not define factor 1. Again, there is evidence of cross loading, with the six items from factor 2 also loading on factor 1. However, Wilson et al. (1991) do not elaborate on the criterion value for factor loading and no loading scores were reported. Overall the evidence suggests a unidimensional structure to the TRQ. This interpretation is consistent with Kennedy et al.'s (2004) analysis of the individual items in the TRQ which found that >75% of items addressed the psychological and emotional effects (see Figure 1).

Wilson et al. (1991) recommend that the separate scores from the two factor solution, not the four factor solution, might be used to explain individual differences in tinnitus distress. Although this recommendation was made, it would appear not to have been followed by subsequent researchers. For example, the TRQ is often interpreted as having four valid subscales derived from the four factor solution (see Kennedy et al., 2004; Newman and Sandridge, 2004; Holgers et al., 2003).

Internal Consistency

The TRQ has also been shown to have excellent internal consistency for all 26 items in different populations ($\alpha = 0.96$) suggesting that all the items are measuring overlapping aspects of psychological distress (Wilson et al., 1991).

Questionnaire Construct

Construct Validity: Convergent Validity

Although there is limited evidence for convergent validity, we conclude that the TRQ appears to have high convergent validity. Robinson et al. (2003) reported that the TRQ highly correlates with the THQ (r = 0.78), the TQ (r = 0.82) and the THI (r = 0.88).

Construct Validity: Discriminant Validity

There is a large body of evidence which, taken overall, indicates that the TRQ has low discriminant validity and hence is sensitive to generalised emotional distress. Wilson et al. (1991) tested the TRQ in three different populations and found that in all three samples the TRQ strongly correlated with anxiety scales: Taylor Manifest anxiety scale (Bendig, 1956; r = 0.66) and Spielberger state-trait anxiety inventory (Spielberger et al., 1970; r = 0.58/60 to r = 0.71/74). Furthermore, that there were extremely strong correlations between the TRQ and the BDI (r = 0.63/87) (Wilson et al., 1991). Strong correlations have also been replicated by Robinson et al. (2003) and Andersson et al. (2003). Robinson et al. (2003) reported that the TRQ strongly correlated with BDI (r = 0.66) and moderately correlated with the Hamilton Rating Scale for Depression (r = 0.52). Andersson et al. (2003) reported strong correlations with the Hospital Anxiety and Depression scales (r = 0.69 for depression and r = 0.72 for anxiety). Robinson et al. (2003) observed that the TRQ weakly correlates with the Private Self-Consciousness Scale (r = 0.20) and moderately correlates with the Quality of Well-Being scale (r = -0.39) and the Modified Somatic Perception Questionnaire (r = 0.53).

Interpretation of the Scores

The TRQ was developed to distinguish levels of tinnitus-related distress. Evidence for the TRQ as a measure of tinnitus severity is limited. Although, Wilson et al. (1991) claim that the TRQ is a useful screening questionnaire that can distinguish tinnitus sufferers, no grading system for categorising tinnitus severity has been explicitly developed.

Sensitivity to Change

Despite claims to effectively evaluate the effects of psychological interventions on tinnitus, Wilson et al. (1991) do not provide any evidence for this psychometric property. For example, no research has been conducted to find a minimal clinically important change score. Nevertheless the TRQ is consistently used as evidence for the efficacy of interventions.

Wilson et al. (1991) did provide evidence of a high test-retest reliability (r = 0.88) (over a retest period of 3 days to 3 weeks), suggesting that the TRQ is stable across time. However, without a minimal clinically important change score, the interpretation of any change in TRQ scores pre- and post-treatment is difficult. For example, there is no way of knowing whether a change in score of 2 points or 5 points is needed for the change to be considered meaningful.

Summary

In conclusion, the information available on the development and structure of the TRQ is unclear. The factorial structure and internal consistency seem to point towards a unidimensional structure focusing on the emotional impact of tinnitus, and the low discriminant validity indicates that this is generalised to the same psychological domains as depression and anxiety. Clinical interpretations of the TRQ scores are limited by the fact that no grading system was developed for quantifying tinnitus severity and no minimal clinically important change score has been recommended.

Tinnitus Handicap Inventory (THI)

The 25-item Tinnitus Handicap Inventory is reported to be a brief diagnostic and screening tool that measures the impact of tinnitus on everyday function (Newman et al., 1996). The THI consists of three response options; yes (4 points), sometimes (2 points) and no (0 points). The total score is rescaled from the weighted sum of all the items so that the global score ranges from 0 – 100, with a higher score denoting increased levels of tinnitus distress (Table 3). The THI was originally designed as a companion to the Hearing Handicap Inventory for Elderly (HHIE, Ventry and Weinstein, 1982) and the

Dizziness Handicap Inventory (DHI, Jacobson and Newman, 1990) to complete a set of tools to quantify perceived handicap in a variety of hearing related diseases, such as Ménière's disease (Newman and Sandridge, 2004).

Initial Questionnaire Development

Again, limited information makes it difficult to ascertain the exact process behind the questionnaire development (Newman et al., 1996). Clinical expertise certainly played a role. For example, Newman stated: "The alpha-THI consisted of 45-items derived empirically from case histories of patients with tinnitus" (Newman et al., 1996, p. 144). Although, the authors infer that a combination of content and face validity were used to source and develop the items, the information provided is unclear. Some items were adapted from the HHIE and the DHI, and also from Tyler and Baker (1983; see Fig 2). The THI has adopted the identical response options and scoring as the DHI and HHIE. No further information was provided on the item development and content.

The 45 items were reduced to 25 items (final version) using response frequency distributions, item-total correlations, and content validity. Items were eliminated if high endorsement rates for one response option were found, or if item-total correlations were $r \leq 0.50$. Newman et al. (1996) believed these items were insensitive, discriminated little between subjects, and did not represent the scale concept. There is little clarity on who conducted the content validity, which items were removed and how the criteria were decided for the frequency distribution elimination score (i.e., 85% patients selecting same response) and for the item-total correlation scores ($r \leq 0.50$).

Questionnaire Structure

Factorial Structure

Following the examination of the item content, a three-domain model for the THI was proposed (Newman et al., 1996) to cover limitations in (i) functioning (12 items: mental, social and physical), (ii) emotional response to tinnitus (8 items), and (iii) the desperation associated with tinnitus (5 items: catastrophic). However, the domain content (subscales) were solely based on content validity and the questionnaire structure was not subjected to empirical validation at the time.

In 2003, Baguley and Andersson investigated the theoretical factorial structure and three domains originally proposed by Newman et al. (1996). The THI scores were predefined into the three factor solution and subjected to PCA. Initial item-total correlations were rather high (r = 0.60) for a scale which assumes to be measuring a broad construct. This result suggests that the THI is tapping only into narrow characteristics (Clark and Watson, 1995). In contrast to the structure previously proposed, oblique rotation revealed that the majority of items loaded onto the one factor reflecting functioning (19 items). Factor 2 (emotional) and factor 3 (catastrophic) each consisted of 5 items, some of which also loaded onto factor 1. The high overlap between factors suggests that they are not distinctive from each other and are in fact measuring the same latent variables (Zachariae et al., 2000). This unidimensional structure is further evident in the THI Danish (Zachariae et al., 2000) and Italian translations (Monzani et al., 2008), and in Kennedy et al.'s (2004) analysis of tinnitus questionnaires where the THI was shown to favour psychological/emotional distress (68%; Figure 1). Therefore, despite Newman's initial claim of a three-factor structure, the THI appears to be rather unidimensional.

Internal Consistency

During the initial development process, the only analysis conducted was Cronbach's alpha (Newman et al., 1996). This measure alone does not provide a sufficient measure of the internal structure and its reliability. Validation of the factorial structure is required too (Streiner, 2003; Cortina, 1993). Excellent Cronbach's alpha scores were reported for the THI 25-items (α = 0.93) suggesting that all the items measure the same underlying construct. High scores such as this indicate redundancy of items and could be indicative of the unidimensional structure, although internal consistency is not sufficient evidence for homogeneity (Green et al., 1977; Schmitt, 1996). Both the functional and emotional subscales showed high internal consistency (α = 0.86 and α = 0.87 respectively). The internal consistency for the catastrophic subscale was questionable (α = 0.68), potentially due to the small number of items in this scale. Coefficient alpha values are vulnerable to the number of test items (Schmitt, 1996). Recent evidence suggests that the questionable Cronbach's alpha score for catastrophic subscale is more likely to indicate a heterogeneous construct with poor inter-relatedness between the items (Tavakol and Dennick, 2011).

Questionnaire Construct

Construct Validity: Convergent Validity

Overall, the THI has high convergent validity demonstrating that it measures a construct that is comparable to other tinnitus questionnaires. The THI shows strong correlations with the THQ (r = 0.78), the symptom rating scales (r = 0.67 - 0.72) (Newman et al., 1996), the TQ (Spearman's r = 0.89) (Baguley et al., 2000) and the TRQ (r = 0.89) (Robinson et al., 2003).

Construct Validity: Discriminant Validity

Overall, the THI measures a construct that is independent of generalised emotional distress. Moderate discriminant validity for the THI has been observed by the moderate correlations with the BDI (r = 0.32 - 0.58), the Hamilton Rating Scale for Depression (r = 0.49) and the Quality of Well-Being scale (r = -0.37) (Newman et al., 1996; Robinson et al., 2003). Weak to moderate correlation with the Modified Somatic Perception Questionnaire scores (r = 0.24 - 0.38) and the Private Self-Consciousness Scale (r = 0.23) (Robinson et al., 2003).

Interpretation of the Scores

The THI was developed to grade tinnitus severity at intake assessment. Developed by Newman et al. (1998) using the quartiles analysis on test-retest reliability data, the grading system originally defined four categories (Table 5). These categories were further developed by a UK working group (McCombe et al., 2001). The authors recommended adopting a five category grading system as a way to diagnose and screen tinnitus severity in clinical practice (Table 5). This recommendation is based on the expert opinions and knowledge. For example, "[…] take the known epidemiological and clinical data and with our own clinical, medico-legal and research experience to produce a severity scale" (McCombe et al., 2001, p.26). No further evidence has been reported for the development of these categories other than a reference that the scores were based on Newman et al.'s 1998 analysis. In turn, no empirical evidence has been provided on the validity of the definitions. This prompts questions about the reliability of these categories to provide clinicians with valid meanings behind the scores.

Sensitivity to Change

The THI was not developed for use as an outcome measurement tool and does not explicitly maximise the responsiveness to treatment-related effects (Meikle et al., 2007). Nevertheless, it is used internationally as an outcome measure for testing the effectiveness of therapeutic interventions and for clinical research. Newman et al. (1998) conducted test-retest reliability to assess stability over short time intervals. The THI showed high test-retest reliability ($r = 0.92$) suggesting that a mean score would unlikely deviate more than 7.0 points (based on SEM) between tests (Newman et al., 1998).

The variance around these scores (test and retest) produced the 95% confidence intervals. From examining the 95% confidence interval and the SEM, Newman et al. (1998) concluded that a reduction of ≥20 points is required for the change to be classified as clinically meaningful (Newman et al., 1998). However, this reduction is dependent on the intake assessment score being >20 points, therefore potentially failing to observe the small changes that occur in patients with slight or mild tinnitus distress (Table 5).

In 2011, Zeman et al. investigated the minimal clinically important change required to show a noticeable improvement in the THI scores. They used an anchor-based technique; CGI-I rating with seven response options (Table 6a). Again, patient groups were formed according to the CGI-I scores (four groups: much better, minimally better, no change, worse). The effect size (Cohen d) separating the minimally better and no change groups ($d = 0.5$) was used alongside the estimated standard deviation of the before-after difference ($SD = 14$) to calculate the minimum significant change for improvement ($\Delta THI = 0.7$). Therefore, patients perceived a THI score reduction of 7 points as a meaningful improvement. This offers a more sensitive measure of responsiveness in the THI compared to the previous scores suggested by Newman et al. (1998), as it takes into account patient experience.

Despite these attempts to produce a criterion for a significant change score, the THI will always receive criticism for its inherent lack of sensitivity in the three response options.

Summary

In conclusion, the THI may not be as robust as first proposed. There is little clarity provided on the item development and it has a unidimensional structure which emphasises the emotional aspects of tinnitus distress. A grading system for quantifying tinnitus severity has been developed for the THI and does provide useful clinical interpretation. Finally, there have been attempts to provide a minimal clinically important change score.

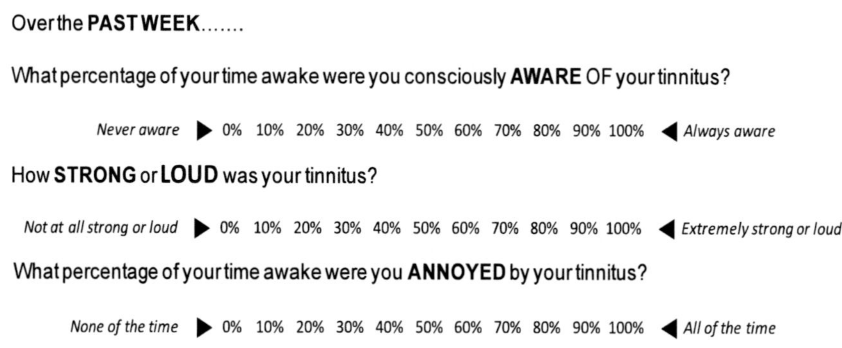

Figure 4. Example of the response options format on the Tinnitus Functional Index. Adapted from Meikle et al. (2012).

Tinnitus Functional Index (TFI)

The 25-item TFI was a large international collaboration of 21 investigators (including 17 expert judges). It was specifically developed to be (i) discriminative to provide measures of tinnitus distress, (ii) evaluative to provide a responsive measure of treatment-related changes, and (iii) comprehensive to cover multiple domains of tinnitus severity. For each item, individuals are asked to rate their judgement on a scale of 0 -10. Each item scale has descriptive anchors at either end that vary depending on the item content (Figure 4). The total score is rescaled from the weighted sum of the items in each subscale so that the global score ranges from 0 to 100, with a higher score denoting higher levels of distress (Table 3).

Initial Questionnaire Development

The report of the development is extremely detailed. Meikle et al. (2012) reports the development of the TFI through prototype 1 and 2 to the final 25-item questionnaire. Initially, 175 items were sourced from nine widely used tinnitus questionnaires excluding ambiguous items (i.e., referring to multiple subtopics) and overly negative items. The 175 items were assigned to 13 different domains of tinnitus handicap, on the consensus of all judges. Duplicate items were reduced to one. Judges completed a three-point responsiveness rating scale which also provided the basis for item selection. Prototypes 1 and 2 were administered on four occasions at five or four sites respectively. Both prototypes were completed by over 300 participants (the first occasion). However it should be noted that there was a high dropout rate after this first administration. The majority of the participants were recruited from Veterans' Affairs (VA) hospitals. This is potentially problematic because VA patients are not representative of the wider clinical population. The majority of VA patients tend to be male and to experience severe tinnitus with co-morbidities.

Forty-three items were used to create prototype 1 with a 0 – 10 response scale. Thirteen items were removed after examination of response frequency distributions, effect size and exploratory factor analysis. Eight items were removed because the effect size for individuals whose tinnitus was unchanged or worse (based on global question) was unacceptably large (i.e., 0.89) and therefore considered unresponsive to perceived change. A variety of the factorial structure techniques were used on prototype 1 data to inform item choice and questionnaire structure. PCA followed by Principal Axis Factoring and oblique rotations revealed eight factors (8 of the proposed 13 domains were identified). A predefined 'somatic' domain was removed due to low factor loading for two items, one of which also showed floor effects (below the desired lower limit of 2.0). The remaining three items were removed due to low factor loadings or initial mean score. The criterion for low scores was not reported. Thirty items remained after this analysis to create prototype 2. Although the same examinations were conducted on prototype 2, none of these findings appear to contribute to the decision to remove the final five items.

This decision was based on "careful examination" of the items (content validity). The final version was a 25-item TFI with eight subscales (Table 3).

Questionnaire Structure

During the development of the TFI, reliability and validity testing (i.e., factor analysis, effect size and convergent/discriminant validity) were conducted on prototypes 1 and 2. The final 25-item TFI has not yet been subjected to a formal statistical validation. So far, evidence for its validation comes from a re-analysis of the relevant subset of data collected using the 30-item prototype 2. In this section, all the evidence provided is based on this re-analysis.

Factorial Structure

The TFI covers eight different domains of tinnitus distress. The 30-item prototype was subjected to PCA followed by Principal Axis Factoring, with orthogonal and oblique rotations informing both model's factor extraction. The eight factor solution revealed in prototype 1 was also clearly revealed for the 30-item prototype 2, despite the reduction in items.

To validate the factorial structure, the 25-item prototype 2 data was again subjected to Principal Axis Factoring, with oblique rotations. The same eight factors were clearly identified; (i) cognition (3 items), (ii) auditory (3 items), (iii) intrusiveness (3 items), (iv) sleep (3 items), (v) relaxation (3 items), (vi) QoL (4 items), (vii) emotional (3 items), and (viii) sense of control (3 items). Only one item (TFI Q21) is shown to load on more than one factor (i.e., Cognitive and Quality of Life). The remaining 24 items load onto a single factor, demonstrated by the high loading values.

Internal Consistency

The 25-item prototype 2 was shown to have an excellent Cronbach's alpha score for all 25 items ($\alpha = 0.97$) and for the eight subscales (range from $\alpha = 0.82 - 0.97$). This score suggests that all the items in the questionnaire are measuring a similar construct. The reliable factorial structure indicates that the items are measuring different aspects of tinnitus distress, rather than the presence of redundant items.

Questionnaire Construct

Construct Validity: Convergent Validity

There is limited evidence for convergent validity for the TFI. Presently, there is only the evidence from the 25-item prototype 2 data. Meikle et al. (2012) reported extremely strong correlations between the 25-item prototype 2 and the THI ($r = 0.86$), i.e., high convergent validity. Further evidence is needed.

Construct Validity: Discriminant Validity

Again, there is limited evidence. The 25-item prototype 2 moderately correlates with the BDI-primary care ($r = 0.56$), demonstrating moderate discriminant validity with an independent measure of emotional distress.

Interpretation of the Scores

Meikle et al. (2012) did propose a grading system based on preliminary data analysis using the 25-item prototype 2 data (Table 5). Throughout development, the five categories (not a problem to a very big problem) from the global severity question (i.e., how much of a problem is your tinnitus?) have been compared to the mean scores from prototype 1 and 2. A strong association between these scores has been shown throughout ($p < 0.001$). Following these comparisons, the mean 25-item prototype 2 scores were categorised based on the individual responses to the global severity question. The response distributions of the total scores in these categories were examined to produce the grading system (Table 5).

Sensitivity to Change

The TFI was developed to effectively evaluate treatment-related change and this was explicitly considered at all steps of its design (Meikle et al., 2012). The items were judged on responsiveness by a panel of judges. The 0 – 10 response options allow for small changes in the scores to be detected.

The 25-item prototype 2 data demonstrated high test-retest reliability for the overall 25-item scores (r = 0.76) (retest period of 7 to 30 days). For the eight subscales, high test-retest reliability has been observed with the scores ranging from 0.66 to 0.90 (Meikle et al., 2012). Therefore, the 25-item prototype 2 and the subscales show stability over short time intervals.

Based on 25-item prototype 2 data, Meikle et al. (2012) did propose a minimal clinically important change score. The mean scores were categorised into five groups based on global perception of change scores at 3 and 6 months (Table 6c). Having examined the data in the plot for differences in scores between the much-to-moderately changed group and the unchanged at 3 and 6 months, Meikle et al. proposed a reduction of around 13 points to be meaningful to patients (Figure 5).

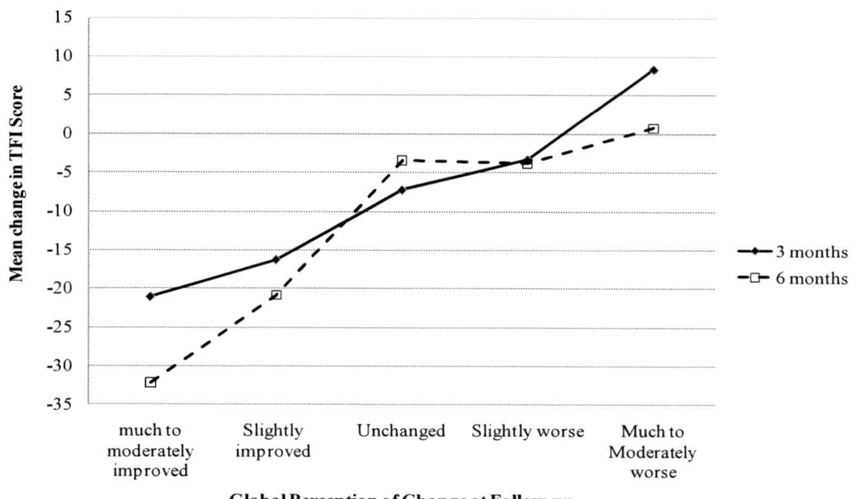

Figure 5. Overall mean TFI change scores at 3 and 6 month follow-ups after grouping patients based on Global Perception of Change score. A minimal clinically important change score of -13 points was determined based on the difference between the unchanged and much-to-moderately improved groups at 3 and 6 months. Adapted from Meikle et al. (2012).

Summary

The development process of the TFI is clearly described and the authors provide clear, in-depth information on all of the steps that were followed. The supplemental published information provides useful evidence from the different prototypes. So far the evidence would suggest that the TFI covers eight different domains of tinnitus distress, is comparable to at least one other tinnitus questionnaire and is measuring more than just generalised emotional distress. The authors have provided both a grading system and a minimal clinically important change score, therefore providing clinical interpretations to the TFI scores. However, this evidence is all based on preliminary data analysis with a rather limited sample of participants. Despite the rigorous validation conducted during the development of the TFI, the final TFI has not been subjected to formal validation.

Conclusion

Our review seeks to promote quality assurance in validation research concerning tinnitus questionnaires. Although beyond the scope of this book chapter, we recommend that the 'gold standard' would be to carry out a systematic review of the literature before selecting any given tinnitus questionnaire for a service audit or clinical trial. This would provide a comprehensive and critical overview of the existing questionnaires available. In addition to the measurement properties, selection might also give consideration to the suitability of the tinnitus questionnaire for the study population, the potential burden of completing the tinnitus questionnaire (e.g., length, question difficulty, emotional impact of certain questions), and the practical aspects (e.g., copyright costs, complexity of scoring method). We make the assumption that the questionnaire validation is conducted on a sample of people with tinnitus that is representative of the population in which the instrument is to be used.

Explicit quality criteria for studies on the development and evaluation of tinnitus questionnaires help to legitimise recommendations about what is the 'best' questionnaire for tinnitus handicap. A wide range of resources are available to help the researcher develop questionnaire instruments and evaluate existing tools. We draw the reader's attention to one useful web resource that provides practical guidance on how to select, translate and validate questionnaires (http://www.emgo.nl/kc/about.html), with an electronic quality assurance handbook available in English (see also Terwee et al., 2007). Following these guidelines, in Table 7, we summarise the evidence reviewed and reported in this chapter according to ten quality criteria for good measurement properties, arranged under the four key categories (see also Uijen et al., 2012). According to these criteria, the TFI comes out "best", meeting eight of the ten criteria according to the current published development reports. However, one caveat to this recommendation is that the final 25-item version of the TFI has not been subjected to these formal validation criteria. This should be the subject of future research.

Table 7. Summary of the critical evaluation of the psychometric properties of the five tinnitus questionnaires

Questionnaire	Content validity	Internal consistency	Structural validity	Construct validity		Reproducibility		Responsiveness	Floor or ceiling effects	Interpretability
				Convergent	Discriminant	Agreement	Reliability			
Tinnitus Questionnaire	?	++	?	?	?	0	?	+	0	?
Tinnitus Handicap Questionnaire	+–	++	+	+	+	?	+	0	?	?
Tinnitus Reaction Questionnaire	?	+–	?	+	?	0	+	0	?	0
Tinnitus Handicap Inventory	?	++	+	+	+	?	+	+	?	+
Tinnitus Functional Index	++	+–	+	+	+	0	+	+	–	+

Rating: + = positive, 0 = no information available, – = poor, ? = indeterminate rating (doubtful design or method), where two symbols are given this indicates that several different validation criteria were used.

References

Adamchic, I., Tass, P. A., Langguth, B., Hauptmann, C., Koller, M., Schecklmann, M., Zeman, F., and Landgrebe, M. (2012). Linking the Tinnitus Questionnaire and the subjective Clinical Global Impression: Which differences are clinically important? *Health and Quality of Life Outcomes,* 10 (1),79

Andersson, G., Baguley, D.M., McKenna, L., and McFerran, D. (2005). *Tinnitus: A multidisciplinary approach*. London: Whurr Publishers Ltd.

Andersson, G., Kaldo-Sandström, V., Ström, L., & Strömgren, T. (2003). Internet administration of the Hospital Anxiety and Depression Scale in a sample of tinnitus patients. *Journal of psychosomatic research*, 55(3), 259-262.

Baguley, D.M., and Andersson, G. (2003). Factor Analysis of the Tinnitus Handicap Inventory. *American Journal of Audiology*, 12, 31-34.

Baguley, D.M., Humphriss, R.L., and Hodson, C.A. (2000). Convergent validity of the tinnitus handicap inventory and the tinnitus questionnaire. *The journal of Laryngology and Otology*, 114, 840-843.

Beck, A.T., Steer, R.A., Ball, R., Ciervo, C.A., and Kabat, M. (1997) Use of the Beck Anxiety and Depression Inventories for primary care with medical outpatients. *Assessment*, 4, 211–219.

Bendig, A. W. (1956). The development of a short form of the Manifest Anxiety Scale. *Journal of Consulting Psychology*, 20, 384.

Clark, L. A., and Watson, D. (1995). Constructing Validity: Basic Issues in Objective Scale Development. *Psychological Assessment*, 7(2), 309-319.

Cortina, J. M. (1993). What is Coefficient Alpha? An Examination of Theory and Applications. *Journal of Applied Psychology*, 78(1), 98-104.

Cronbach, L. J. (1951). Coefficient Alpha and the Internal Structure of Tests. *Psychometrika*, 16(3), 297-334.

De Vet, H.C., Terwee, C.B., Ostelo, R.W., Beckerman, H. Knol, D.L., and Bouter, L.M. (2006). Minimal changes in health status questionnaires: distinction between minimally detectable change and minimally important change. *Health and Quality of Life Outcome*, 4: 54

Department of Health (2009). *Provision of Services for Adults with Tinnitus. A Good Practice Guide.* London: Central Office of Information.

Devon, H.A., Block, M.E., Moyle-Wright, P., Ernst, D.M., Hayden, S.J., Lazzara, D.J., Savoy, S.M., and Kostas-Polston, E. (2007). A Psychometric Toolbox for Testing Validity and Reliabilty. *Journal of Nursing Scholarship*, 39(2), 155-164.

Fenigstein, A., Scheier, M.F., and Buss, A.H. (1975). Public and private self-consciousness: Assessment and theory. *Journal of Consulting and Clinical Psychology*, 43, 522-527.

Field, A. (2009) *Discovering Statistics using SPSS* (3rd Ed). London: Sage Publications Ltd.

FItzpatrick, A.R. (1983). The Meaning of Content Validity. *Applied Psychological Measurement*, 7(1), 3-13.

Floyd, F. J. and Widaman, K. F. (1995). Factor Analysis in the Development and Refinement of Clinical Assessment Instruments. *Psychological Assessment*, 7(3), 286-299.

Gerhards, F., Brehmer, D., and Etzkorn, M. (2004). Dimensionalität des Tinnitus-Fragebogens. *Verhaltenstherapie*, 14, 265-271.

Goebel, G. and Hiller, W. (1998) *Tinnitus-Fragebogen (TF).* Ein Instrument zur Erfassung von Belastung und Schweregrad bei Tinnitus. Handanweisung. Göttingen: Hogrefe

Goebel, G., Kahl, M., Arnold, W., and Fichter, M. (2006). 15-year prospective follow-up study of behavioural therapy in a large sample of inpatients with chronic tinnitus. *Acta Oto-Laryngologica*, 126, 70-79.

Green, S.B., Lissitz, R.W., and Mulaik, S.A. (1977). Limitations of coefficient alpha as an index of test unidimensionality. *Educational and Psychological Measurement*, 37, 827—838.

Hallam, R.S. (1996). *Manual of the Tinnitus Questionnaire (TQ)*. London: The Psychological Corporation.

Hallam, R.S. (2008). *TQ Manual of the tinnitus questionnaire: Revised and updated.* London: Polpresa Press.

Hallam, R.S., Jakes, S.C., & Hinchcliffe, R. (1988). Cognitive variables in tinnitus annoyance. *British Journal of Clinical Psychology*, 27, 213-222.

Hamilton, M. (1960). A rating scale for depression. *Journal of Neurology, Neurosurgery, and Psychiatry*, 23, 52-62.

Haynes, S.N., Richard, D.C.S., and Kubany, E.S. (1995). Content Validity in Psychological Assessment: A Functional Approach to Concepts and Methods. *Psychological Assessment*, 7(3), 238-274.

Henry, J.L., & Wilson, P.H. (1998). The Psychometric Properties of Two Measures of Tinnitus Complaints and Handicap. *International Tinnitus Journal*, 4(2); 114-121.

Hesser, H. (2000). Methodological considerations in treatment evaluations of tinnitus distress: A call for guidelines. *Journal of Psychosomatic Research*, 69, 305-307.

Hiller, W., & Goebel, G. (1992) A psychometric study of complaints in chronic tinnitus. *Journal of Psychosomatic Research*, 36(4), 337-348.

Hiller, W., and Goebel, G. (2004). Rapid assessment of tinnitus-related psychological distress using the mini-TQ. *International Journal of Audiology*. 43; 600-604

Hiller, W., Goebel, G., and Rief, W. (1994) Reliability of self-rated tinnitus distress and association with psychological symptom pattern. *British Journal of Clinical Research*, 33, 231-239.

Hoare, D.J., Gander, P.E., Collins, L., Smith, S., and Hall, D.A. (2012) Management of tinnitus in English NHS audiology departments: an evaluation of current practice. *Journal of Evaluation in Clinical Practice*, 18, 326-334.

Holgers, K. M., Barrenas, M. L., Svedlund, J., and Zoger, S. (2003). Clinical Evaluation of Tinnitus: A Review. *Audiological Medicine*, 2, 101-106.

Ireland, C. E., Wilson, P. H., Tonkin, J. P., and Platt-Hepworth, S. (1985). An evaluation of relaxation training in the treatment of tinnitus. *Behaviour Research and Therapy,* 23, 423-430.

Jacobson, G.P. and Newman, C.W. (1990). The development of the Dizziness Handicap Inventory. *Archives of Otolaryngology - Head and Neck Surgery*, 116, 424-427.

Jacobson, N. S., Follette, W. C., and Revenstorf, D.(1986) Toward a Standard Definition of Clinically Significant Change. *Behavior therapy*, 17(3), 308-311.

Jacobson, N. S., Roberts, L. J., Berns, S. B., and McGlinchey, J. B. (1999). Methods for Defining and Determining the Clinical Significance of Treatment Effects: Description, Application, and Alternatives. *Journal of Counselling and Clinical Psychology,* 67(3), 300-307.

Jakes, S. C., Hallam, R. S., Chambers, C and Hinchcliffe, R. (1985). A factor analytical Study of Tinnitus Complaint Behaviour. *Audiology*, 24; 195-206.

Kam, A.C., Cheung, A.P., Chan, P.Y., Leung, E.K., Wong, T.K., Tong, M.C., and Van Hasselt, A. (2009). Psychometric properties of a Chinese (Cantonese) version of the Tinnitus Questionnaire. *International Journal of Audiology*, 48(8), 568-575.

Kaplan, R.M., Ganiats, T.G., and Seiber, W.J. (1996). *The Quality of Well Being Scale, Self-Administered (QWBS-SA).* San Diego: University of California-San Diego.

Kennedy, V., Wilson, C., and Stephens, D. (2004). Quality of Life and Tinnitus. *Audiological Medicine*, 2; 29-40.

Kirshner, B., and Guyatt, G. (1995). A methodological framework for assessing health indices. *Journal of Chronic Disorder*, 38, 27-86.

Kuk, F. K., Tyler, R. S., Russell, D., and Jordan, H. (1990). The Psychometric Properties of a Tinnitus Handicap Questionnaire. *Ear and Hearing*, 11(6), 434-445.

Lohmann, N. (1980). A factor analysis of life satisfaction, adjustment and morale measures with elderly adults. *The International Journal of Aging and Human Development,* 11(1), 35-43.

Main, C.J. (1983) The Modified Somatic Perception Questionnaire (MSPQ). *Journal of Psychosomatic Research*, 27, 503-514.

McCombe, A., Baguley, D., Coles, R., McKenna, L., McKinney, C., and Windle-Taylor, P. (2001). Guidelines for the grading of tinnitus severity: the results of a working group commissioned by the British Association of Otolaryngologists, Head and Neck Surgeons, 1999. *Clinical Otolaryngology*, 26, 388-393.

Meeus, O., Blaivie, C., and Van De Heyning, P. (2007). Validation of the Dutch and the French version of the Tinnitus Questionnaire. *Royal Belgian Society for Ear, Nose, Throat, Head and Neck surgery (B-ENT),* 3(7); 11-17.

Meikle, M. B., Henry, J. A., Griest, S. E., Stewart, B. J., Abrams, H. B., McArdle, R., Myers, P. J., Newman, C. W…and Vernon, J. A. (2012) The Tinnitus Functional Index: Development of a New Clinical Measure for Chronic, Intrusive Tinnitus. *Ear and Hearing*, 33(2), 153-176.

Meikle, M. B., Stewart, B. J., Griest, S. E., and Henry, J. A. (2008). Tinnitus Outcomes Assessment. *Trends in Amplification*, 12(3), 223-235.

Meikle, M.B., Stewart, B.J., Griest, S.E., Martin, W.H., Henry, J.A., Abrams, H.B., McArdle, R., Newman, C.W., and Sandridge, S.A. (2007). Assessment of tinnitus: Measurement of treatment outcomes. *Progress in Brain Research*, 166, 511-521.

Monzani, D., Genovese, E., Marrara, A., Gherpelli, C., Pingani, L. Forghieri, M., Rigatelli, M., Guadagnin, T., and Arslan, E. (2008). Validity of the Italian adaptation of the Tinnitus Handicap Inventory; focus on quality of life and psychological distress in tinnitus-sufferers. *Acta Otorhinolaryngologica Italica,* 28, 126-134.

Newman, C. W., Wharton, J. A., and Jacobson, G. P. (1995). Retest stability of the tinnitus handicap questionnaire. *The Annals of Otology, Rhinology and Laryngology,* 104(9/1); 718-723.

Newman, C.W., and Sandridge, S.A. (2004). Tinnitus Questionnaires. In: SNOW Jr, J.B. *Tinnitus: Theory and Management.* Ontario: BC Decker Inc, pp. 237-254.

Newman, C.W., Jacobson, G.P., and Spitzer, J.B. (1996). Development of the Tinnitus Handicap Inventory. *Archives of Otolaryngology - Head and Neck Surgery,* 122, 143–8.

Newman, C.W., Sandridge, S.A., and Jacobson, G.P. (1998). Psychometric adequacy of the Tinnitus Handicap Inventory (THI) for evaluating treatment outcome. *Journal of the American Academy of Audiology*, 9, 153–60.

Newman, C.W., Sandridge, S.A., Bea, S.M., Cherian, K., Cherian, N., Kahn, K.M., and Kaltenbach, J. (2011). Tinnitus: Patients do not have to 'just live with it'. *Cleveland Clinic Journal of Medicine*, 78(5), 312-319.

NHS White Paper (2010). Department of Health, Equity and excellence: Liberating the NHS, London, TSO, 2010. Available at: http://www.dh.gov.uk/en/Publicationsandstatistics/Publications/PublicationsPolicyAndGuidance/DH_117353

Parkerson Jr, G. R., Gehlbach, S. H., Wagner, E. H., James, S. A., Clapp, N. E., and Muhlbaier, L. H. (1981). The Duke-UNC Health Profile: an adult health status instrument for primary care. *Medical care*, 806-828.

Peterson, R.A. (1994). A Meta-analysis of Cronbach's Coefficient Alpha. *Journal of Consumer Research*, 21, 381-391.

Revicki, D, Hays, R.D., Cella, D., AND Sloan, J. (2008). Recommended methods for determining responsiveness and minimally important differences for patient-reported outcomes. *Journal of Clinical Epidemiology*, 61, 102-109.

Robinson, S.K., McQuaid, J.R., Viirre, E.S., Betzig, L.L., Miller, D.L., Bailey, K.A., Harris, J.P., and Perry, W. (2003). Relationship of Tinnitus Questionnaires to Depressive Symptoms, Quality of Well-Being, and Internal Focus. *The International Tinnitus Journal*, 9(2), 97-103.

Schmitt, N. (1996) Uses and Abuses of Coefficient Alpha. *Psychological Assessment*, 8(4), 350-353.

Seydel, C., Zirke, N., Haupt, H., Szczepek, A., Olze, H., and Mazurek, B. (2012) Psychometric instruments for the diagnosis of tinnitus. *HNO-Praxis*, 60; 732-742.

Spielberger, C. D., Gorsuch, R. L., & Lushene, R. E. (1970). *STAI manual.* Palo Alto, CA: Consulting Psychologists Press.

Stevens, C., Walker, G., Boyer, M., and Gallagher, M. (2007) Severe tinnitus and its effect on selective and divided attention. *International Journal of Audiology,* 46, 208-216.

Streiner, D.L. (2003) Starting at the Beginning: An Introduction to Coefficient Alpha and Internal Consistency. *Journal of Personality Assessment*, 80(1), 99–103.

Streiner, D.L. and Norman, G.R. (2008) *Health measurement scales: a practical guide to their development and use.* (4th Ed) New York: Oxford University Press.

Sullivan, M., Katon, W, Russo, J., Dobie, R., and Sakai, C. (1993) A randomized trial of nortriptyline for severe chronic tinnitus. Effects on depression, disability, and tinnitus symptoms. *Archives of Internal Medicine*, 11, 2251–2259

Tavakol, M. and Dennick, R. (2011) Making sense of Cronbach's alpha. *International Journal of Medical Education*, 2, 53-55

Terwee, C. B., Bot, S. D.M., de Boer, M. R., van der Windt, D. A.W.M., Knol, D. L., Dekker, J., Bouter, L. M., and de Vet, H. C.W. (2007). Quality criteria were proposed for measurement properties of health status questionnaires. *Journal of Clinical Epidemiology*, 60, 34-42.

Tyler, R.S., and Baker, L.J. (1983) Difficulties experienced by tinnitus sufferers. *Journal of Speech and Hearing Disorders,* 48, 150–4.

Uijen, A. A., Heinst, C. W., Schellevis, F. G., van den Bosch, W. J., van de Laar, F. A., Terwee, C. B., and Schers, H. J. (2012) Measurement Properties of Questionnaires Measuring Continuity of Care: A Systematic Review. *PLoS ONE* 7(7): e42256. doi:10.1371/journal.pone.0042256

Ventry, I.M., and Weinstein, B.E. (1982) The Hearing Handicap Inventory for the Elderly: a new tool. *Ear and Hearing*, 3, 128–34.

Wilson, P. H., and Henry, J. L. (1998). Tinnitus Cognitions Questionnaire: development and psychometric properties of a measure of dysfunctional cognitions associated with tinnitus. *The international tinnitus journal*, 4(1), 23.

Wilson, P.H., Henry, J., Bowen, M., and Haralambous, G. (1991) Tinnitus Reaction Questionnaire: Psychometric Properties of a Measure of Distress Associated with Tinnitus. *Journal of Speech and Hearing Research*, 34, 197-201.

Zachariae, R., Mirz, F., Johansen, L.V., Anderson, S.E., Bjerring, P., and Pedersen, C.B. (2000) Reliability and validity of a Danish adaptation of the Tinnitus Handicap Inventory. *Scandinavian Audiology*, 29, 37–43.

Zeman, F., Koller, M., Figueiredo, R., Aazevedo, A., Rates, M., Coelho, C., Kleinjung, T., De Ridder, D., Langguth, B., and Landgrebe, M. (2011) Tinnitus handicap inventory for evaluating treatment effects: which changes are clinically relevant? *Otolaryngology-Head and Neck Surgery*, 145, 282–287.

Zung, W. (1965) A self-rating depression scale. *Archives of General Psychiatry*, 12, 63-70.

In: Tinnitus
Editors: F. Signorelli and F. Turjman

ISBN: 978-1-63117-556-5

Chapter 3

Mindfulness Based Tinnitus Stress Reduction: A New Treatment with Ancient Roots

Jennifer J. Gans* and Michael Cole
[1]Tinnitus Practitioner's Association (TPA), Hamden, CT, US
Assistant Clinical Professor, Department of Otolaryngology,
University of California, San Francisco, CA, US
[2]Research Scientist, VA Northern California Health Care System
Assistant Clinical Professor, Department of Psychology,
University of California, Berkeley, CA, US
Associate Clinical Professor, Department of Neurology,
University of California, Davis, CA, US

Abstract

This chapter aims to report on a novel mind–body intervention, Mindfulness Based Tinnitus Stress Reduction (MBTSR), as a beneficial treatment for chronic tinnitus. Mindfulness-based approaches have been scientifically investigated for treating a range of diseases and symptoms over the past several decades. Recent research has begun to look into this approach for managing the symptom of tinnitus. This chapter will review

* E-mail: jg@MindfulTinnitusRelief.com. www.MindfulTinnitusRelief.com.

this research, and look more closely at a recent pilot study conducted by the author. In this pilot study, eight tinnitus patients at the University of California, San Francisco (UCSF) Audiology Clinic participated in an MBTSR pilot study. The program included 8 weeks of group instruction on mindfulness practice, a 1-day retreat, supplementary readings, and home-based practice using meditation CDs. Using a pre–/post-intervention design, mean differences (paired t-tests) were calculated. Benefits were measured by a reduction in clinical symptoms, if present, and a tinnitus symptom perception shift. Tinnitus symptom activity and discomfort as well as psychological outcomes were assessed by self-report questionnaires. Both quantitative and qualitative data were gathered. Results indicate that Effect Sizes may be clinically significant and demonstrate a substantial decrease for items measuring perceived annoyance and perception of handicap of tinnitus. Change scores on study measures all moved in the hypothesized direction, with the exception of negligible change found for the Acting with Awareness factor of mindfulness. This pilot study addresses some of the limitations of previous research and provides preliminary evidence that an 8-week MBTSR program may be an effective intervention for treating chronic tinnitus and its comorbid symptoms, and may help reduce depression and phobic anxiety while improving social functioning and overall mental health. These promising findings warrant further investigation with a randomized controlled trial. The online version of the program, MindfulTinnitusRelief.com, is discussed.

Keywords: Mindfulness, Tinnitus, Stress, Management, MBTSR, MBSR, MindfulTinnitusRelief.com, Mindfulness based tinnitus stress reduction· Mindfulness based stress reduction

Introduction

While many theories have been proposed to explain the occurrence of tinnitus, it is a multimodal disorder that can have different causes and different pathophysiologies. This makes tinnitus difficult to treat, and historically interventions have met with only variable success (Meikle et al., 2007).

In an extensive review of the literature from 1980 to 2009, Holmes and Padgham (2009) reported on the impact tinnitus can have on a person's life. For those who experience tinnitus bothersome enough to consult their doctor, tinnitus is most commonly associated with symptoms of anxiety, sleep disturbance, and depression (Andersson et al., 2005; Lockwood et al., 2002). Disrupted sleep is the most significant complaint (Megwalu et al., 2006),

medical or surgical treatment is effective; likewise, there is a wide range of psychological effects, and lifestyle and general health are particularly affected. Perceived lack of control over the symptom, problems with aspects of attention and focus, maladaptive coping strategies, catastrophic thinking, as well as the use of similar methods of treatment (e.g., cognitive coping strategies, CBT, and relaxation techniques) are common features of both chronic pain and tinnitus (Møller 2000; Tonndorf 1987). One of the first studies looking at mindfulness as a means of pain management was conducted by Kabat-Zinn, Lipworth, and Burney (1985). They studied 90 patients with chronic pain and trained them in a 10-week stress reduction and relaxation program. Results demonstrated significant reductions in present moment pain, negative body image, inhibition of activity by pain, and mood disturbances such as anxiety and depression. Significant improvements in activity level and feelings of self-esteem were also reported. These reductions and improvements, with the exception of present moment pain, were maintained 15 months post- treatment for all subjects. These studies demonstrated promise for mindfulness-based interventions to successfully address a chronic, afflictive condition in which other treatments have had limited to no success, such as is the case in tinnitus.

A few studies have recently begun to look at meditation-based treatments for chronic tinnitus (Philippot et al., 2011; Sadlier et al., 2008). In a pilot study conducted in Wales by Sadlier et al., (2008), a combination of CBT and mindfulness training over four 1-hour sessions was used to treat 25 individuals with chronic tinnitus. Split into two groups, the treatment group received CBT and Mindfulness Meditation practice, while the second group (control group) waited 3 months and was then treated with the same intervention. Significant reductions in tinnitus severity post-treatment were reported following the CBT and Mindfulness Meditation practice intervention, with 80% of patients reporting that they were better or much better at 4- and 6-month follow-ups. This study provides important initial evidence of a mindfulness-based intervention to potentially represent an important component of an efficacious treatment package to address tinnitus, yet it remains unclear as to the extent mindfulness practice specifically contributed to symptom reduction.

A second randomized controlled trial conducted by Philippot et al., (2011) examined the efficacy of mindfulness training alone as compared to relaxation training in 25 individuals with tinnitus. All subjects were first offered a single session of psychoeducation regarding tinnitus, followed by six weekly sessions of either mindfulness or relaxation training with each session lasting 2 hours and 15 minutes. Results indicated that psychoeducation, followed by mindfulness training, appeared to be an effective intervention for chronic tinnitus with a reduction in negative emotions, rumination and psychological difficulties from living with tinnitus. These benefits were enhanced or at least maintained by mindfulness training over time, as compared to relaxation training, where there was an erosion of these benefits over the follow-up.

Though the above two studies by Philippot and colleagues and one by Sadlier and colleagues provide encouraging evidence for use of mindfulness-based interventions to treat tinnitus, it is of interest to determine the efficacy of a tinnitus treatment that directly applies mindfulness techniques to working with experienced tinnitus symptoms. The recent pilot study described in more detail later in this chapter does exactly that, whereby it aims to test the feasibility of a novel mind–body intervention, Mindfulness Based Tinnitus Stress Reduction (MBTSR), as a possible tinnitus intervention, to address limitations mentioned in previous studies, and to determine whether conducting a larger randomized controlled trial is warranted. This particular intervention seeks to generate a tinnitus perception shift in its participants. For the purposes of this study, a tinnitus perception shift is defined as a person's attitude towards the reality of living with chronic tinnitus changing from one of struggle to one of acceptance. Furthermore, although mindfulness-based training has been used with tinnitus patients in recent studies, the present study is the first to introduce the 8-week MBTSR program with a 7-h day-long mindfulness retreat based on conjecture that longer training would provide participants the full benefits of the intervention (Philippot et al., 2011; Sadlier et al., 2008).

Suppression vs. Acceptance Strategies for Managing Tinnitus

A key aspect of tinnitus treatment, which is particularly relevant to development and application of mindfulness-based interventions, is that of suppression versus acceptance approaches for symptoms reduction. Of all ways to manage chronic tinnitus, suppression of the stimulus is usually the first approach tried. Suppression can be defined here as the intentional inhibition, ignoring, or exclusion of unacceptable thoughts, emotions, or sensations. "Just don't pay attention to it" or "Find something to take your mind off of it" are phrases commonly heard from friends and professionals. When the environment is filled with adequate stimulation, some people with tinnitus can distract, ignore, or suppress the distress, anxiety or other disruptive feelings brought on by tinnitus. They can suppress the tinnitus signal with sound therapy devices, by thinking of something else, or involving themselves in an attention-consuming activity. However, these very same people during times of low stimulation, may find their tinnitus taking center stage. It is at these times that a person's best efforts to suppress, ignore and distract may fail.

The fact that people with tinnitus often complain that their tinnitus is most bothersome when they are in a quiet environment with little to distract, is perhaps one reason why sleep difficulty is one of the most common complaints of people with tinnitus. Around bedtime, when all is quiet and stimulation is low, people with tinnitus report being up for hours with ineffective ways of distracting themselves from the ringing.

An alternative management strategy is acceptance. Acceptance here refers to the welcoming of thoughts, emotions, and other sensations in the moment, with non-evaluative awareness. Mindfulness-Based approaches focus on applying acceptance strategies to work with chronic tinnitus in situations where tinnitus bother is not readily relieved by suppression techniques. Programs like MBTSR suggests an alternative strategy that addresses the failure of suppression strategies to bring relief during times of quiet and low stimulation.

The comparative effectiveness literature on the use of suppression or acceptance techniques to manage chronic pain, another persistent disorder, may inform our understanding of suppression and acceptance's roles as a tinnitus management strategy. In the context of the treatment of pain, the majority of experimental studies have shown acceptance strategies to be more effective at increasing pain tolerance than were other pain-regulation strategies, such as distraction or suppression (Gutierrez et al., 2007; Masedo et al., 2004; McMullen et al., 2008). In one meta-analysis, results showed suppression of pain stimuli is associated with psychopathology such as anxiety and depression (Aldao, Nolen-Hoeksema, and Schweizer 2010). In a randomized controlled trial comparing both acceptance and suppression of pain, results showed similar reductions in pain for both the acceptance and suppression groups; while acceptance led to greater reductions in anticipatory anxiety than suppression (Braams et al., 2011).

The tinnitus research has similar findings. A growing body of research suggests that, when compared to approaches such as distraction, suppression, and cognitive avoidance, acceptance and mindfulness-based strategies appear to be a more adaptive increasing a person's tolerance of tinnitus while reducing negative affect (Westin, Ostergren, and Andersson, 2008; Westin et al., 2011; Westin et al., 2008). It seems that attending mindfully to the tinnitus sensation rather than suppressing, may be a better approach to reduce distress and facilitate adaptive responding to the tinnitus sensation.

Both suppression and acceptance may be effective strategies for different people, to different degrees, under different circumstances, at different times. For the people who consider their tinnitus to be non-bothersome, they likely have had success with applying distraction strategies to manage discomfort. This, however, is not the case for all people with tinnitus all of the time. For those experiencing sleep difficulty, anxiety and depression, suppression techniques have likely failed. At this point it seems helpful to try acceptance strategies, as taught and practiced in MBTSR, to work with the unpleasant tinnitus sensation.

People new to meditation sometimes say, "I'm so bad at meditating. My thoughts are always distracting me." But mindfulness is not about suppressing thoughts. Rather, mindfulness encourages a non-elaborative experiencing of whatever comes in to one's awareness at any given moment. The wandering mind – with its intricate flow of thought processes or "mindstream" – can take us very far away from our actual moment-to-moment experience. Furthermore, this unconscious drifting often leads to storytelling and rumination about things secondary to our actual experience. Once this is recognized, these thoughts are not pushed away, rejected, or suppressed, but rather are seen as objects worthy of note or observation in that moment. But once this secondary elaboration is consciously acknowledged, we self-direct our attention back to the wonder of this moment. For example, one practice that can be helpful in cultivating mindfulness is 'noting.' When a meditator finds herself worrying about how she will possibly get through the day tomorrow after a sleepless night, or feels regret about all the ways her life has been affected by tinnitus, instead of being swept away by these thoughts, she can gently bracket these experiences as "planning" or "sadness", then try to return to an awareness of what is actually happening in *this* moment. This practice can help meditators to acknowledge thoughts, feelings, and experiences without becoming consumed by them.

People with tinnitus might at first reject the idea of exploring and relating differently and more kindly to an unpleasant sensation such as tinnitus that they have tried so hard to push out of consciousness. But if suppression, ignoring and distraction do not always work, acceptance-based strategies, such as MBTSR, may be indicated. The 8-week course teaches the participant to bring awareness to whatever is in their field of awareness at any given moment. Why? Simply because it is already there. Few could argue that it is in the present moment where life actually happens. The past has already passed and the future is yet a best guess on what may or may not come to be. It is only in the present moment when we can choose to respond in new and different ways. MBTSR encourages us to bring awareness to the present experience with an orientation of acceptance, compassion, openness, and curiosity. With this special awareness, a meditator can approach even the most unpleasant experience with a stance of, "This is what is happening in this moment, so let me feel it." When a person brings awareness to whatever the moment brings, be it a pleasant, unpleasant, or neutral (being neither pleasant nor unpleasant), one finds that one can live through it. As stated by Jon Kabat-Zinn regarding using this mindfulness approach: "You can't stop the waves, but you can learn to surf."

Attention and Concentration

There is a finite amount of stimuli that our human mind can pay attention to in any given moment. For this reason we must be judicious with what we pay attention to and for how long. Many say that tinnitus makes concentration on even simple tasks very challenging. For example, obsessive/anxious/angry thoughts can consume one's attentional stores leaving little else for the observation or space needed for responding in new potentially more adaptive ways. Building the flexibility to consciously bring attention to where you want it to go can increase attention and is a skill that people with bothersome tinnitus would likely benefit from strengthening. For people with tinnitus, their awareness is brought to the unpleasant sensation a disproportionate amount of the time.

Many complain that their tinnitus is all that they can think about. Even at times when it is not perceived as active, people's attention may be consumed by anxious thoughts of the next time it will be there. Shifting one's attention to sensations, thoughts, and emotions occurring in the moment, without ruminating on the past or worrying about the future, allows for more effective dealing with life stressors such as tinnitus that may otherwise lead to feelings of anxiety, depression, and sleep difficulty.

What Can You Do When Distraction Doesn't Work?

Of all ways to manage chronic tinnitus, distraction or ignoring the sensation when it is bothersome and intrusive is usually the first tried. As noted previously, "Just don't pay attention to it" or "Find something to take your mind off of it" are phrases often heard from friends and professionals. Distraction and ignoring techniques are taught as component of some clinically effective treatments such as Cognitive Behavioral Therapy (CBT) and Tinnitus Retraining Therapy (TRT). But what about those times when a person's best efforts to ignore and distract fail? This commonly happens to people at night when all is quiet and few compelling distractions are available. It is not uncommon for people with chronic tinnitus to develop sleep disorders as a result. It is in these quiet moments that the mind is prone to ruminating on past events, catastrophizing about a foreseeable future, and the like.

Programs like Mindfulness Based Tinnitus Stress Reduction (MBTSR) take tinnitus management to the next step, where other treatments fall short. MBTSR teaches skills addressing the times when ignoring and distraction just don't work. MBTSR does not promote distraction and ignoring but, in fact, teaches the opposite skill of moving into the bare sensation. This "moving into" or awareness is a very unique and special kind of attention to all experiences moment by moment. This special kind of attention is accompanied by the qualities of curiosity, openness, acceptance, non-judgment, and compassion to whatever is in a person's field of awareness, even unpleasant sensation such as tinnitus is perceived.

People with tinnitus might at first reject the idea of exploring and relating differently and more kindly to an unpleasant sensation that they have tried so hard to push out of consciousness. But if ignoring and distracting does not work, this seems like a feasible option to try. The 8-week course teaches the participant to bring awareness to whatever is in their field of awareness at any given moment. Why? Because it is there. When a person brings awareness to whatever the moment brings, be it a pleasant, unpleasant, or neither pleasant nor unpleasant, one finds that they can, in fact, live through it. It is the stories that we tell ourselves about what may or may not happen in the future that keeps us running from past to the future without stopping in the present, where life actually happens. MBTSR and other programs teaching and practicing acceptance in relation to unpleasant sensations such as tinnitus can be effective management tools for anyone experiencing bothersome tinnitus as well as for those with tinnitus who have tried other approaches with little relief.

Taking the High Road and the Low Road

It is important to also consider stress in relation to tinnitus, and how the stress reduction component of mindfulness-based interventions lends itself to tinnitus symptom reduction. First and foremost, our brains are wired to maintain our survival. Often, dangers require us to react so quickly that the relatively time-consuming act of evaluating the threat is a luxury we cannot afford. As Joseph LeDouix of NYU has introduced to us, when we are faced with a real or perceived threat, our body has two simultaneous processes going on. He has coined these processes the "High Road" and the "Low Road." The Low Road is the quick-and-dirty, must-act-now reaction where we take in information through our senses and send it directly to the thalamus. The thalamus is a relay station where sensory information is then distributed. The thalamus sorts out the different sensory information and its meaning is appraised as either important or not important. If a threat seems to be present, the amygdala then sends an alert to the hypothalamus, which then triggers our sympathetic nervous system to take action. Subsequently, stress hormones such as epinephrine and cortisol are released from the adrenal gland, setting the fight-or-flight response in motion.

Simultaneously, the "High Road" process is unfolding. Here, our sensory cortex takes in information. It is within the sensory cortex that we can consciously see that an incoming stimulus can be perceived in a myriad of ways. It looks at all of the options and references these options based on stored memories in the hippocampus that either support or negate a particular interpretation of a sensation. If, through a conscious observation of the stimuli, the sensation is determined to be non-threatening, regions of the frontal cortex including the orbital frontal cortex and medial frontal cortex exert emotional control whereby activation of limbic regions (e.g., amygdala) is dampened, thus forestalling the fight-or-flight response and restoring a sense of emotional balance.

How does this relate to tinnitus? The tinnitus sensation, while never pleasant, in reality, poses no threat to our survival – yet it is often misconstrued as an imminent danger, setting off the automatic, unconscious, sub-cortical, and reactive "Low Road." The tinnitus signal is appraised by the amygdala to be something that requires hypervigilant alertness. This triggers the hypothalamus to set the sympathetic nervous system into action. The body is submerged in the fight-or-flight response, building up stress in the body, which in turn only worsens the experience of tinnitus.

The question is, how can we calm the amygdala and see the tinnitus signal as a sensation that is benign, albeit unpleasant, but not one that should trigger alarm? By taking the "High Road" and engaging the higher cortical regions (such as the orbital and medial prefrontal cortices) the fear response can be modulated and extinguish the misappraised, subcortically-driven fear reaction. When we consciously bring a sensation like tinnitus to the "High Road," we make room for our reasoning, insight, and fear modulation ability to restore emotional balance and extinguish the fear reaction in exchange for a more accurate, balanced response.

Research conducted by Sara Lazar and her colleagues at Harvard showed that experienced meditators had more cortical growth in the medial prefrontal cortex and thickening in the right insula. In her study, the amount of medial prefrontal growth was in direct proportion to the participant's number of years of experience with meditating. Perhaps training in mindfulness meditation such as in an 8-week training program like MBTSR can enlist the medial prefrontal cortex to help extinguish the automatic fear reaction to a non-threatening stimulus like tinnitus. Relaxing overly alert attention and letting go of negative appraisal of tinnitus may open up new possibilities for learning new ways to live with tinnitus.

Mindfulness Based Tinnitus Stress Reduction Pilot Study: A Symptom Perception Shift Program

Taking into account a couple of the critical considerations for optimizing tinnitus intervention detailed above, specifically utilizing a mindfulness-based intervention that (1) teaches techniques directly applied to management and acceptance of tinnitus symptoms and (2) works directly with the reduction stress so as to minimize the amplifying effects stress can have on tinnitus, a pilot study by Jennifer Gans and colleagues adapted the MBSR protocol to be applied directly to treatment of tinnitus. This new intervention, Mindfulness-Based Tinnitus and Stress Reduction (MBTSR), includes most of the components of MBSR with the addition of modules that train the individuals in mindfulness techniques that are directly applied to tinnitus symptoms as well as modules that provide psychoeducation on tinnitus.

Intervention

While taught in a similar 8-week format as MBSR (Kabat-Zinn 1982, 2005), the MBTSR course includes significant programmatic modifications designed to address participants' unique experience with tinnitus. The most fundamental change was the reshaping and reorientation of MBSR into a program whose central focus addresses psychoeducation related to the often co-occurring disorders (e.g., sleep disorder, anxiety, depression) common in people with tinnitus. Additional class time was focused on guided mindfulness practices emphasizing awareness of sound and tinnitus perception, education about sleep and sleep hygiene, explaining the circular and complex connection between stress in daily life and tinnitus exacerbation, increasing relaxation and awareness skills, and increasing an overall sense of well-being as it relates to living with tinnitus. The additional class time was added to improve the participants' understanding of the relationship between tinnitus and the comorbid symptoms. The program also aims to foster the recognition of early warning signs of tinnitus exacerbation, teaching skills to come back to present, moment-by-moment awareness whenever the mind starts to dwell in the past or future, and teaching how to access inner resources through the acquisition of mindfulness skills. In class and home-practice mindfulness exercises emphasized becoming aware of the tinnitus sensation, observing it with a certain spaciousness and affectionate curiosity instead of reacting in habitual ways. Participants were instructed to practice mindfulness throughout their day when tinnitus is perceived to be loud and troubling, including while eating meals, before sleeping, during social interactions, and during periods of quiet.

Similar to the original MBSR program, the MBTSR course introduces participants to mindfulness practice in the form of sitting meditation, body awareness, mindful movement, as well as integrating informal mindfulness practices into activities of daily life (e.g., eating, walking, washing one's hands). In accordance with the non-goal setting that is inherent in mindfulness, participants were encouraged to let go of decreasing tinnitus as the "goal" of the program. Instead, participants were encouraged to develop a mindful outlook on their lives as a whole. Between sessions, participants were asked to practice at home for 30 min/day, 6 days/week, aided by meditation CDs made by the course instructor. At the start of the program, each participant was supplied with a copy of *Full Catastrophe Living* (Kabat-Zinn, 2005), and a participant manual. Between classes, participants were instructed to enhance their participation through readings from the course materials.

Participant and instructor manuals were created specifically for the study. The MBTSR manual was derived from Kabat-Zinn's (1982, 2005) MBSR curriculum, with weekly sessions following a similar format, with classes lasting 150 minutes. Classes began with a 40-minute meditation practice, followed by a review of the previous week's homework, presentation and discussion of the theme of that particular session, and concluding with a brief experiential exercise and a review of the homework for the following week. The seventh and eighth sessions were slightly modified from the standard session format; in order to emphasize self-reliance in the meditative practice, participants were given fewer instructions and less guidance during exercises. Similar to MBSR, between the sixth and seventh weeks of the program, participants attended a 7-hour retreat. During the daylong retreat, participants experienced various meditations (i.e., awareness of the breath, sitting, walking meditations) bringing moment-to-moment awareness to body sensations, thoughts, and emotions as they arise. The retreat provided participants with the opportunity to practice the skills they had cultivated over the past 6 weeks, with the support of a continuous group setting. The day ended with a group discussion of participants' individual experience with the practice, where they were encouraged to reflect on what they had learned.

At the end of the eighth and final MBTSR class, participants were asked to anonymously complete a feedback form developed for the study. Finally, participants were provided with a packet of post- assessment measures along with a self-addressed stamped envelope, and were asked to mail the completed forms back to the Clinic. Pre- and post-outcome data was received and included in analysis from all eight study participants.

Efficacy of MBTSR

Using a pre-post intervention design, benefits of MBTSR was evaluated in eight subjects by measuring a reduction in clinical symptoms, if present, and a tinnitus symptom perception shift. Both quantitative and qualitative data were gathered. Pre-assessment measures were individually administered to participants immediately after consents were signed, no earlier than 2 weeks prior to the start of the intervention. Quantitative results indicated that there was moderate to large improvement with respect to reduced tinnitus annoyance and awareness, increased mindfulness (specifically for non-judging of inner experience or refraining from evaluations of one's cognitions, sensations, and emotions), reduction in mood disturbance (primarily for depression, phobic anxiety, and somatization), and meaningful improvements in patient well-being and quality of life. A comparison of the quantitative and qualitative data demonstrates the cross-validity of participants' responses regarding certain aspects of functioning, including enhanced well-being and reduced depression, anxiety, somatization, and sleep difficulty.

Qualitative findings also suggest that within this small group there was a reduction in the need for sleep medication, change in how participants experienced and accepted their bodies, and a marked shift in their perception of tinnitus. Across the board, subjects expressed that they had a positive experience with the intervention, and noticed a positive change in their day-to-day lives.

It is not clear whether the mechanisms for positive changes can be better accounted for by the psychological benefits of participating in the group rather than any reduction in tinnitus distress, per se. One participant mentioned that they gained significant insight into their experience with tinnitus by "just being in the group discussing a problem that everyone is sharing." This comment may point to the fact that non-specific effects of the treatment, such as group support, may have influenced treatment benefit. The attention paid to participants by the group instructor may also have had an influence on treatment results. It will be important for future studies of MBTSR to include an active control group to account for possible group effects and the effects of interactions with a group facilitator.

Observed findings in the data, if duplicated in a larger study, would provide significant subjective benefit to people suffering with tinnitus after participating in an 8-week MBTSR training program. The current study adds to the literature by exploring a mindfulness-based intervention that is specifically focused on managing tinnitus symptoms as well as stress. Should further research confirm the current findings, MBTSR is likely to be an effective adjunct treatment for chronic tinnitus.

Conclusion and Future Directions

Initial evidence for the efficacy of mindfulness-based interventions to improve symptoms associated with tinnitus has been provided by a small handful of studies to date. Results from the study of Sadlier et al., (2008) indicated significant reductions in tinnitus variables using a combination treatment of CBT and meditation offering four 1-hour sessions to participants. Using a 6-week program, Philippot et al., (2011) found that training in mindfulness may be a useful addition to psychoeducation about tinnitus. Work by Gans and colleagues (2013) in developing the Mindfulness-Based Tinnitus and Stress Reduction (MBTSR) follows more closely the 8-week MBSR program including a full day of mindfulness practice (Kabat-Zinn 1982, 2005) while also for the first time tailoring mindfulness techniques to directly address management of tinnitus symptoms and stress. Findings from this pilot study suggest that MBTSR was effective in reducing tinnitus associated symptoms as well as frequently comorbid symptoms.

Although the above described studies are promising, it will be important to validate the findings in larger, double-blinded, randomized controlled studies. Further, each of these studies includes components that are in addition to mindfulness techniques that could serve as nonspecific factors for the intervention (e.g., cognitive-behavioral interventions, psychoeducation, group therapy) precipitating, at least in part, the beneficial effects on tinnitus symptoms. Thus, future studies can investigate each of these individual components as compared to the fully integrated interventions so as to determine the extent to which each nonspecific factor contributes to the beneficial effects. This would provide more definitive evidence that the mindfulness training component of the intervention is indeed precipitating beneficial effects in and of itself. Further, this information regarding nonspecific components will allow interventions to be better tuned so as to eliminate surplus content while also affording the possibility of bolstering particularly effective content. Another area of important follow-up will be investigating intervention program length. As previously noted, participants in the Philippot et al., study felt that the 6-week program was too short. Varying program length will be important so as to maximize symptom improvement balanced with not structuring program length too long so as to risk significant participant dropout, burnout, etc.

Given the consistent positive effects found thus far for mindfulness-based interventions to address tinnitus associated symptoms, it is worth considering deploying this type of intervention for tinnitus in formats in addition to the group setting. This might be particularly effective in advanced technological settings such as telemedicine, use of web-based mediums, and smartphone and computer tablet applications. One such example is the newly developed web-based intervention for the MBTSR course (MindfulTinnitusRelief.com). This online course is taught by the founder and lead researcher of MBTSR. Similar to the MBTSR course used in research and presented in this chapter, the 8-week online course teaches the skills to live *with* instead of *against* tinnitus, especially for those of whom suppression of the sensation has not brought adequate relief. Including online modules accompanied by audio and video recordings, interactive discussions, and home-practice assignments, this online course reflects the particular approach and skills taught in the MBTSR course and is now available for use.

This type of online learning makes it possible to serve the millions of people who have not been able to benefit from other tinnitus management options. The intelligent applications of online learning such as this make it possible to serve more people with fewer resources and relieves hospital and community settings from the budgetary pressures that can be a roadblock to care such as human resources, classroom space, materials, and labs. Furthermore, with these technological approaches, treatment of tinnitus associated symptoms can be accessed in rural regions where services might not be otherwise available.

Another important area of follow-up will be that of investigating the neural substrates of the changes precipitated by mindfulness-based interventions in tinnitus. Identification of the neural systems involved with the beneficial effects of mindfulness training will allow for greater fine-tuning of the interventions by maximizing the targeting of these systems. For instance, if the functioning of regions associated with the stress response are found to be strongly associated with changes in perceived averseness of tinnitus, then interventions might be made more efficacious by enhancing the stress reduction content of the intervention. Alternatively, if neural systems involved in cognitive and/or emotional control are strengthened by the intervention and are associated with better ability to manage tinnitus symptoms, then the focus on content areas relevant to cognitive and/or emotional control can be bolstered in interventions so as to improve their efficacy. In addition, identifying neural systems associated with changes in tinnitus symptoms will shed greater light on the neural mechanisms involved with producing tinnitus, which to date is poorly understood.

This chapter suggests that there may be a sub-classification of tinnitus patients that can be divided into those who benefit from suppression and distraction strategies, those who benefit from a combination of suppression and acceptance strategies, and those who benefit from acceptance strategies. Perhaps a sub-classification of tinnitus patients may improve therapeutic allocation, which, in turn, may improve therapeutic success for individual patients. Future research may look at this potential sub-classification.

In the development of tinnitus knowledge, research, treatment, and understanding, there is an ongoing search to find ways to manage this chronic symptom. We hope that studies will continue to explore the range of healing that can take place by encouraging patients to learn the skills needed to help them change their relationship to tinnitus.

Conflict of Interests

The author's state that there is no conflict of interest including financial or other personal considerations having the potential to compromise or bias professional judgment and objectivity in this chapter.

References

Andersson (2005). Tinnitus: a multidisciplinary approach. London: Whurr.

Beck (1979). Cognitive therapy of depression. New York: Guilford.

Beddoe, A., & Murphy, S. (2004). Does mindfulness decrease stress and foster empathy among nursing students? The Journal of Nursing Education, 43(7), 305–Brown, K. W., & Ryan, R. M. (2003). The benefits of being present: mindfulness and its role in psychological well-being. *Journal of Personality and Social Psychology*, 84(4), 822–848.

Caldwell (2010). Developing mindfulness in college students through movement based classes: effects on mood, self-regulatory self- efficacy, stress, and sleep quality. *Journal of American College Health*, 58, 433–442.

Carlson, L. E., & Garland, S. N. (2005). Impact of mindfulness-based stress reduction (MBSR) on sleep, mood, stress and fatigue symptoms in cancer outpatients. *International Journal of Behavioral Medicine*, 12, 278–285.

Carlson, L. E., Speca (2004). Mindfulness-based stress reduction in relation to quality of life, mood, symptoms of stress and levels of cortisol, dehydroepian- drosterone sulfate (DHEAS) and melatonin in breast and prostate cancer outpatients. *Psychoneuroendocrinology,* 29(4), 448–474.

Chiesa, A., & Serretti, A. (2009). Mindfulness based cognitive therapy for psychiatric disorders: a systematic review and meta-analysis. *Psychiatry Research,* 187, 441–453.

Cohen, J. (1977). Statistical power analysis for the behavioral sciences (2nd ed.). New York: Academic Press.

Davis, C. G., & Morgan, M. S. (2008). Finding meaning, perceiving growth, and acceptance of tinnitus. *Rehabilitation Psychology*, 53, 128–138.

Department of Veterans Affairs, Veterans Benefits Administration, *Annual Benefits Report*, FY 2010,5.

Folmer, R. L. (2002). Long-term reductions in tinnitus severity. *Bio- medical Central Ear, Nose and Throat Disorders*, 2, 3.

Goleman, D. (1988). The meditative mind: The varieties of meditative experience. New York: Putnam.

Grossman U. (2007). Mindfulness training as an intervention for fibromyalgia: evi- dence of postintervention and 3-year follow-up benefits in well-being. *Psychotherapy and Psychosomatics*, 76(4), 226–233.

Grossman, P., Kappos, L., Mohr, D. C., Gensicke, H., D'Souza, M., Penner, I. K., et al., (2010). MS quality of life, depression and fatigue improve after mindfulness training: a randomized trial. *Neurology*, 75, 1141–1149.

Gutierrez (2004). Comparison between an acceptance-based and a cognitive-control-based protocol for coping with pain. *Behavioral therapy*, 35:767-783.

Harrop-Griffiths (1987). Chronic tinnitus: association with psychiatric diagnoses. *Journal of Psychosomatic Research*, 31, 613–622.

Hebert, S., & Lupien, S. J. (2007). The sound of stress: blunted cortisol reactivity to psychosocial stress in tinnitus sufferers. *Neuroscience Letters,* 411(2), 138–142.

Heller, A. J. (2003). Classification and epidemiology of tinnitus. *Otolaryngologic Clinics of North America*, 36(2), 239–248.

Holmes, S., & Padgham, N. (2009). Review paper: more than ringing in the wars: a review of tinnitus and its psychosocial impact. *Journal of Clinical Nursing*, 18(21), 2927–2937.

Kabat-Zinn, J. (1982). An outpatient program in behavioral medicine for chronic pain patients based on the practice of mindfulness meditation: theoretical considerations and preliminary results. *General Hospital Psychiatry*, 4, 33–47.

Kabat-Zinn, J. (1994). Wherever you go, there you are. New York: Hyperion.

Kabat-Zinn, J. (2005). Full catastrophe living: Using the wisdom of your body and mind to face stress, pain, and illness: Fifteenth (anniversaryth ed.). New York: Bantam Dell.

Kabat-Zinn, J., Lipworth, L., & Burney, R. (1985). The clinical use of mindfulness meditation for the self-regulation of chronic pain. *Journal of Behavioral Medicine*, 8, 163–190.

Kabat-Zinn, J., Lipworth, L., Burney, R., & Sellers, W. (1987). Four- year follow up of a meditation based program for the self regula- tion of chronic pain: treatment outcomes and compliance. *The Clinical Journal of Pain,* 2, 159–17.

Kabat-Zinn (1992). Effectiveness of a meditation-based stress reduction program in the treatment of anxiety disorders. *The American Journal of Psychiatry,* 149, 936–943.

Kabat-Zinn, J., Wheeler, J. E., Light, T., Skillings, Z., Scharf, M. J., Cropley, T. G., et al., (2003). Part II: Influence of a mindfulness meditation-based stress reduction intervention on rates of skin clearing in patients with moderate to severe psoriasis undergoing phototherapy (UVB) and photochemo-therapy (PUVA). *Con- structivism in the Human Sciences*, 8, 85–106.

Kaplan, K. H., Goldenberg, D. L., & Galvin-Nadeau, M. (1993). The impact of a meditation- based stress reduction program on fibro- myalgia. *General Hospital Psychiatry*, 15(5), 284–289.

Ludwig, D. S., & Kabat-Zinn, J. (2008). Mindfulness in medicine. *Journal of the American Medical Association,* 300, 1350–1352. doi:10.1001/ jama.300.11.1350.

Ma, S. H., & Teasdale, J. D. (2004). Mindfulness-based cognitive therapy for depression: replication and exploration of differential relapse prevention effects. *Journal of Consulting and Clinical Psychology*, 72, 31–40.

Majumdar, M., Grossman, P., Dietz-Waschkowski, B., Kersig, S., & Walach, H. (2002). Does mindfulness meditation contribute to health? Outcome evaluation of a German sample. *Journal of Alternative and Complementary Medicine, 8(6), 719.*

Martinez Devesa, P., Perera, R., Theodoulou, M., & Waddell, A. (2010). Cognitive behavioural therapy for tinnitus. *Cochrane Database of Systematic Reviews*, 9, CD005233.

Masedo AI, Rosa Esteve M (2007). Effects of suppression, acceptance and spontaneous coping on pain tolerance, pain intensity and distress. *Behavior Research Ther* 45:199-209.

Mason, O., & Hargreaves, I. (2001). A qualitative study of mindfulness-based cognitive therapy for depression. *The British Journal of Medical Psychology*, 74(2), 197–212.

Mazurek, B., Stöver, T., Haupt, H., Klapp, B. F., Adli, M., Gross, J., et al., (2010). The significance of stress: its role in the auditory system and the pathogenesis of tinnitus. Hals-, Nasen-, Ohren- Heilkunde, 58, 162–172.

McMullen J, Barnes-Holmes D, Barnes-Holmes Y, Stewart I, Luciano C, Cochrane A (2008) Acceptance versus distraction: Brief instructions, metaphors and exercises in increasing tolerance for self-delivered electric shocks. *Behav Res. Ther.*, 46, 122-129.

Megwalu, U. C., Finnell, J. E., & Picirillo, J. F. (2006). The effects of melatonin on tinnitus and sleep. *Otolaryngology and Head and Neck Surgery*, 134(2), 210–213.

Meikle (1993). Evaluation of tinnitus treatment: methodological aspects. *Journal of Audiological Medicine*, 2, 141–150.

Meikle, M. B., Stewart, B. J., Griest, S. E., Martin, W. H., Henry, J. A., Abrams, H. B., et al., (2007). Assessment of tinnitus: measurement of treatment outcomes. *Progress in Brain Research*, 166, 511– 521.

Miller, J. J., Fletcher, K., & Kabat-Zinn, J. (1995). Three-year follow- up and clinical implications of a mindfulness meditation-based stress reduction intervention in the treatment of anxiety disorders. *General Hospital Psychiatry*, 17, 192–200.

Møller, A. R. (2000). Similarities between severe tinnitus and chronic pain. *Journal of the American Academy of Audiology*, 11, 115– 124.

Morone, N. E., Lynch, C. S., Greco, C. M., Tindle, H. A., & Weiner, D. K. (2008). "I felt like a new person." the effects of mindful- ness meditation on older adults with chronic pain: qualitative narrative analysis of diary entries. *The Journal of Pain*, 9(9), 841–848.

Philippot, P., Nef, F., Clauw, L., de Romrée, M., & Segal, Z. (2011). A randomized controlled trial of mindfulness based cognitive therapy for treating tinnitus. *Clinical Psychology & Psychotherapy*. doi:10.1002/ cpp.756.

Rosenzweig, S., Reibel, D. K., Greeson, J. M., Brainard, G. C., & Hojat, M. (2003). Mindfulness-based stress reduction lowers psychological distress in medical students. *Teaching and Learning in Medicine*, 15, 88–92.

Sadlier, M., Stephens, S., & Kennedy, V. (2008). Tinnitus rehabilitation: a mindfulness meditation cognitive behavioural therapy approach. *The Journal of Laryngology and Otology*, 122(1), 31–37.

Sanchez, L., & Stephens, D. (1997). A tinnitus problem questionnaire in a clinic population. *Ear and Hearing,* 18, 210–217.

Shapiro, S. L., Schwartz, G. E., & Bonner, G. (1998). Effects of mindfulness-based stress reduction on medical and premedical students. *Journal of Behavioral Medicine*, 21(6), 581–599.

Smith, B. W., Ortiz, J. A., Steffen, L. E., Tooley, E. M., Wiggins, K. T., Yeater, E. A., et al., (2011). Mindfulness is associated with fewer PTSD symptoms, depressive symptoms, physical symptoms, and alcohol problems in urban firefighters. *Journal of Consulting and Clinical Psychology*, 79, 613–617.

Speca, M., Carlson, L. E., Mackenzie, M. J., & Angen, M. (2006). Mindfulness-based stress reduction (MBSR) as an intervention for cancer patients. In R. A. Baer (Ed.), Mindfulness-based treatment approaches: Clinician's guide to evidence base and applications (pp. 239–261). CA, Elsevier Academic Press: San Diego.

Surawy, C., Roberts, J., & Silver, A. (2005). The effect of mindfulness training on mood and measures of fatigue, activity, and quality of life in patients with chronic fatigue syndrome on a hospital wait- ing list: a series of exploratory studies. *Behavioural and Cognitive Psychotherapy*, 33, 103–109.

Sweetow, R. (1986). Cognitive aspects of tinnitus patient management. *Ear and Hearing*, 7, 390–396.

Teasdale, J. D., Segal, Z. V., Williams, J. M. G., Ridgeway, V. A., Soulsby, J. M., & Lau, M. A. (2000). Prevention of relapse/ recurrence in major depression by mindfulness-based cognitive therapy. *Journal of Consulting and Clinical Psychology*, 68(4), 615–623.

Teasdale, J. D., Moore, R. G., Hayhurst, H., Pope, M., Williams, S., & Segal, Z. V. (2002). Metacognitive awareness and prevention of relapse in depression: empirical evidence. *Journal of Consulting and Clinical Psychology,* 70(2), 275–287.

Tonndorf, J. (1987). The analogy between tinnitus and pain: A suggestion for a physiological basis of chronic tinnitus. *Hear Re- search,* 28, 271–275.

Weiss, M., Nordlie, J. W., & Siegel, E. P. (2005). Mindfulness-based stress reduction as an adjunct to outpatient psychotherapy. *Psychotherapy and Psychosomatics*, 74(2), 108–112.

Westin, V., Ostergren, R., and Andersson, G. (2008). The effects of acceptance versus thought suppression for dealing with the intrusiveness of tinnitus. *International Journal of Audiology. 47*(2), 112-118.

Westin, V., Hayes, S. C., & Andersson, G. (2008). Is it the sound or your relationship to it? The role of acceptance in predicting tinnitus impact. *Behaviour Research and Therapy 46*(12), 1259-1265.

Westin, V. Z., Schulin, M., Hesser, H., Karlsson, M., Noe, R. Z., Olofsson, U., & Andersson, G. (2011). Acceptance and commitment therapy versus tinnitus retraining therapy in the treatment of tinnitus: a randomized controlled trial. *Behaviour Research and Therapy*, *49*(11), 737-747.

Zigmond, A. S., & Snaith, R. P. (1983). The Hospital Anxiety and Depression Scale. *Acta Psychiatrica Scandinavica*, 67, 361–370.

Zoger, S., Svedlund, J., & Holgers, K. M. (2001). Psychiatric disorders in tinnitus patients without severe hearing impairment: 24 month follow-up of patients at an audiological clinic. *Audiology*, 40,133–140.

In: Tinnitus
Editors: F. Signorelli and F. Turjman
ISBN: 978-1-63117-556-5

Chapter 4

Potential Roles of Na^+ Influx and Src Family Kinases (SFKs) in the Development of Tinnitus and Hearing Loss

Xian-Min Yu *
BenQ Neurological Institute of Nanjing Medical University, Nanjing, P.R. China

Abstract

Tinnitus is a very annoying symptom which can be induced under many conditions such as damage in the nervous system or ear, exposure to loud sound and side effects of certain medications, etc. Most patients with tinnitus have some degree of hearing loss [147]. Further investigations show that tinnitus and hearing loss are often associated with dysfunction of cochlear spiral ganglion neurons (SGNs) [53, 67, 72, 120, 213]. To understand the process underlying the development of tinnitus, this chapter focuses on recent findings regarding the regulation of neural signaling in SGNs.

Cation currents mediated by voltage-gated sodium (Na^+) and potassium (K^+) channels in SGNs are responsible for transduction of auditory signals encoding simple to complex sounds such as music and

* Correspondence: xianmin.yu88@gmail.com.

language [167, 179]. Alterations of voltage-gated cation currents have been found to be one of important factors causing dysfunctions of SGNs [72, 119] and thereby may be related to the occurrence of tinnitus and hearing loss. Recent mechanistic studies have shown that effects of activation of voltage-gated Na^+ channels on neuronal survival and functions can be abolished by the removal of extracellular Na^+ [75, 207, 210, 211]. This finding indicates that Na^+ influx and/or intracellular Na^+ may play critical roles.

Na^+ is the major cation in the extracellular space, and can enter cells through a variety of routes such as ligand- (e.g., glutamate) and voltage-gated cation (e.g., Na^+) channels [146, 207]. Although roles played by Na^+ influx and/or intracellular Na^+ still need to be characterized in both the regulation of SGN survival and the development of tinnitus, it is known that hypoxic stimulations and/or tissue injury cause significant increases in intracellular Na^+ concentration [15-17, 172]. Up-regulation of voltage-gated Na^+ channels increases caspase-3 activity and cell death whereas the down-regulation of the channels reduces Na^+ entry and cell death [16, 207]. Data including both ours and others' have documented that through the regulation of voltage- and/or ligand-gated Na^+ channels, Src family kinases (SFKs) are involved in the regulation of neuronal development and functions (for recent reviews see [81, 94, 170, 193, 209]). It is reported that noise-induced hearing loss can be prevented with application of SFK inhibitors [25, 87]. Our studies document that effects of intracellular Na^+ on neuronal survival and functions are regulated by SFKs [203, 207, 210, 211], and that voltage-gated Na^+, but not K^+, currents in SGNs are regulated by endogenous SFKs [68]. Application of a SFK inhibitor reduces the amplitude of voltage-gated Na^+ current, enhances steady-state inactivation and delays the recovery of the current in SGNs [68]. By contrast, application of a SFK activator potentiates Na^+ current and promotes the voltage-dependent activation [68]. These findings suggest that the excitability and firing behavior of SGNs to sound stimuli may be regulated by SFKs. To understand how tinnitus and hearing loss may be induced, in this chapter, recent findings dealing with potential roles of Na^+ influx and SFKs are discussed.

1. Introduction

The hair cells in the organ of Corti are innervated by cochlear spiral ganglion neurons (SGNs), which form primary auditory afferents and provide the sole auditory input to the brain. During the development, SGNs acquire specific firing futures which precisely express simple to complex sounds such as music and language. According to anatomical connections to inner or outer hair cells, SGNs are classified into two types. Type I SGNs innervate inner hair cells (IHCs) and the type II innervate outer hair cells (OHCs). 95% of SGNs are found to be type I, which represent the principal encoder of auditory signals [52, 167, 168, 176]. Although functions of type II SGNs remain to elucidate, there are data suggesting that the type II SGNs may provide an integrated afferent feedback loop and amplify both cochlear sensitivity and frequency discrimination [48, 100, 187]. To date, understanding how SGNs function remains a central issue in auditory neuroscience.

According to action potential firings recorded in SGNs, a tonotopic distribution of timing-related parameters and a non-monotonic distribution of excitability within the spiral ganglion have been characterized, which corresponds with the functional organization of the peripheral auditory system [51]. The kinetic features of SGNs change systematically along the tonotopic axis whereas neuronal excitability is distributed such that the most sensitive neurons are located in the mid-cochlear region [51]. The tonotopy is found at every levels in the auditory system, from the spiral ganglion, brainstem to the cortex [105]. More detailed studies at molecular and genetic levels have revealed important mechanisms for understanding how the functions of SGNs are organized [11, 206].

Cation currents mediated by voltage-gated sodium (Na^{+}) and potassium (K^{+}) channels in SGNs are responsible for the generation and propagation of action potentials along the auditory nerve [2, 73, 95, 167, 179]. Voltage-gated K^{+} channels are responsible for cell depolarization-activated K^{+} currents. All types of voltage-gated K^{+} current components have been observed in SGNs, which include delayed rectifier current, low voltage-activated current, dendrotoxin-sensitive current as well as transiently activated current [2, 167]. It is found that the delayed rectifier current, which is mediated with activated Kv3.1 and/or Kv3.2 channels, causes cell repolarization after action potential, whereby allowing the rapid firing [2, 167]. Alteration of the dendrotoxin-sensitive current, which is mediated with Kv1.1, Kv1.2, Kv1.4 and/or Kv1.6 channels, may change the resting membrane potential and thereby affect the transient K^{+} current mediated with activated Kv3.4, Kv4.2 and/or Kv4.3 channels [2, 167]. The transient K^{+} current may speed up repolarization, and thereby enhancing firing frequency [99, 167].

Expressions of these K^{+} channels are found to be very similar in type I versus II SGNs except for Kv1.2 channels. Kv1.2 channels can be found in most of type I but only in 50% of type II SGNs [14]. Kv3.1 channel expression in the spiral ganglion shows an apico-basal gradient [2] and is altered with development [99]. Na^{+}-mediated K^{+} channels such as *Slo* gene-encoded K^{+}

channels [23, 24, 59, 150, 212], are widely distributed throughout the nervous system. These channels are found to be involved in both the regulation of after-potential following action potentials [24, 124], and the protection of neurons from hypoxic stimulation [24, 59, 212].

Expressions of voltage-gated Na^+ channels e.g., Na(V)1.1, Na(V)1.2, Na(V)1.6 and Na(V)1.7 channels are found in SGNs [73, 95, 167]. Na(V)1.1 channels exist along processes of SGNs while Na(V)1.6 and Na(V)1.7 channels are mainly located in the cell body region. Co-labeling with peripherin shows that the expression of Na(V)1.6 and Na(V)1.7 channels is higher in type I than that in type II SGNs [73]. Besides the voltage-gated Na^+ channels, the other types of Na^+ channels (e.g., epithelial Na^+ channels) are also found in SGNs [42, 216].

It has been established that glutamate released from hair cells acts as a principle excitatory transmitter in synapses of type I SGNs [9, 58, 79], which activates α-amino-3-hydroxy-5-methyl-4-isoxazolepropionic acid (AMPA) [78, 131, 148, 165] and N-methyl-D-aspartate (NMDA) receptors expressed in these neurons [148, 198].

It has been proposed that tinnitus and hearing loss may be associated with dysfunction of SGNs [53, 67, 72, 120, 213]. A high dose of aspirin or its active ingredient salicylate, which is widely used to treat pain and vascular occlusive diseases, may induce activation of caspase-3 and 9, soma shrinkage, nuclear condensation and fragmentation in SGNs [53, 67]. These indicate undergoing of apoptosis. Quinine is another drug which can induce hearing loss and tinnitus. An experimental investigation shows that application of the drug to cultured SGNs significantly changes action potentials (broadened in duration and reduced in amplitude) without affecting the resting membrane potential or the input resistance [120]. In voltage-clamp recordings it is found that quinine primarily blocks potassium currents in a voltage-dependent manner at more positive levels [120]. At higher concentrations (>20 microM), quinine can also reduce the size of Na^+ currents in a use-dependent manner, while leaving Ca^{2+} currents relatively unaffected [120].

Quantitative PCR examinations show that following noise stimulation which produces a 20 dB hearing threshold elevation and may cause hearing loss, the mRNA expression of Na(V)1.1 and Na(V)1.6 channels in SGNs are respectively reduced by 29% and 56%. In contrast, Na(V)1.7 mRNA expression is increased by 20% [72]. Immunohistochemical examinations show that after noise stimulation there is increased staining for Na(V)1.1 channels along SGN dendrites, and for Na(V)1.7 channels in the cell body region [72]. These findings have provided a line of direct evidence that noise which leads hearing loss and tinnitus, can cause changes in the expression of voltage-gated Na^+ channels in SGNs.

The expression and function of various voltage- and ligand-gated ionic channels in SGNs are also regulated by *brain-derived neurotrophic factor* (BDNF) or Neurotrophin-3 (*NT-3*) through tyrosine-related kinase B (trkB) or C (trkC) [127, 128]. It has been well established that Src family kinases (SFKs) such as Src or Fyn plays critical roles in the signaling pathway of either trkB or trkC [6, 20, 97, 98, 102, 140, 205, 215]. Through the regulation of voltage- and/or ligand-gated Na^+ channels, as well as actions of intracellular Na^+, SFKs are found to be critically involved in the regulation of neuronal development, survival and functions (for recent reviews see [81, 94, 170, 193, 209]. Application of SFK inhibitors can prevent noise-induced hearing loss [25, 87]. To clarify the process underlying the development of tinnitus, this chapter have focused on recent findings dealing with mechanisms of neural signaling in SGNs and discussed potential roles of Na^+ influx and SFKs.

2. Potential Roles of Na^+ Influx in the Development of Tinnitus and Hearing Loss

A large number of data have demonstrated that exposure to intense sound which may result in hearing loss, causes loss of hair cells [90, 180], and/or damage in synapses of SGNs [156, 158]. Further studies show that following the noise stimulation a significant change in the expression of voltage-gated Na^+ channels occurs [72, 73]. These data have strongly suggested that the alteration in functions of voltage-gated Na^+ channels may contribute to elevated hearing thresholds and to the generation of tinnitus and hyperacusis [72, 73]. Consistently, the enhancement of discharge activity mediated by voltage-gated Na^+ channels is found to underlie the ensemble spontaneous activity (ESA) of the cochlear nerve. Changes in the spectral characteristics of the ESA have been observed in humans with tinnitus [136, 173]. Blocking voltage-gated Na^+ channels such as through application of lidocane or lamotrigine has being used to treat tinnitus patients in the clinic [49, 192].

Recent studies have shown that the effects of activation of voltage-gated Na^+ channels on neuronal survival and functions can be abolished by removal extracellular Na^+ [75, 207, 210, 211]. These findings indicate that Na^+ influx and/or intracellular Na^+ may play critical roles in the regulation of neuronal functions. Na^+ is a major cation in extracellular space. Under resting conditions $[Na^+]_i$ in the cell body region of neurons is about 10 mM [210]. Na^+ can enter cells through a variety of routes including permeation through ligand- (e.g., glutamate) and voltage-gated cation channels (e.g., Na^+ channels), uptaking via membrane exchangers and gradient-driven co-transporters [146, 207]. Short burst or tetanic stimulation of afferents that induces synaptic long-term potentiation (LTP) increases $[Na^+]_i$ up to 40 or 100 mM in spines and adjacent dendrites [163, 164].

Following injury, Na^+ can enter cells through Gd^{3+}-sensitive channels or $Na^+/K^+/2Cl^-$ co-transporters in addition to ligand- or voltage-gated Na^+ channels [174]. Accompanying with Na^+, chloride ions (Cl^-) and water enter cells, leading to acute cell swelling and damage [36, 37]. Moreover, via Na^+-H^+ exchange Na^+ entry may cause pH changes in cells, and thereby alter enzyme activity and cell life [19, 29, 141, 142, 174, 177]. The inhibition of Na^+-H^+ exchange significantly attenuates ischemia-induced cell death [137, 197].

Na^+ influx plays an important role in the onset of anti-Fas-induced apoptosis [30]. Enhancing Na^+ entry by application of the voltage-gated Na^+ channel activator, veratridine, induces caspase-3 activation and cell death [16, 17]. With increases in $[Na^+]_i$ of 10 – 30 mM the agonist bindings to μ-, δ- and κ-types of opioid receptors may be reduced by approximately 60%, 70% and 20%, respectively [201]. The co-occurrence of Na,K-ATPase dysfunction and Na^+ entry may cause AMPAR proteolysis, which leads a rapid reduction of AMPAR cell-surface expression [214]. It has been reported that blocking Na^+ influx protects the nervous tissue from traumatic injuries [3, 4, 17, 46, 65, 85, 183, 185]. Application of Na^+ channel blockers can reduce the sensitization associated with pain [10, 44, 54, 199] and prevent seizures during kindling development [162].

Na^+ entry may also cause an increase in cytosolic Ca^{2+} through either Na^+/Ca^{2+} exchangers or activation of voltage-gated Ca^{2+} channels [27, 107], and thereby activate Ca^{2+}-dependent signaling mechanisms. Excessive intracellular Ca^{2+} is toxic in neurons including SGNs [32, 61, 89, 200, 217]. Under resting conditions $[Ca^{2+}]_i$ in neurons is tightly regulated and normally maintained at 10 – 100 nM. $[Ca^{2+}]_i$ can be altered with Ca^{2+} entry through Ca^{2+} channels (including ligand- and voltage-gated Ca^{2+} channels and non-selective cation channels) located on the plasma membrane. $[Ca^{2+}]_i$ can also be increased through calcium-induced calcium release (CICR) from intracellular Ca^{2+} stores following the activation of inositol trisphosphate receptors (IP3Rs) and/or Ryanodine receptors (RyRs). Using a combination of reverse transcription-polymerase chain reaction, immunolabeling techniques, and confocal Ca^{2+} imaging, all three RyR isoform mRNA transcripts are found in the cell body region of SGNs [118]. RyR3 labeling extends to the synaptic terminals innervating hair cells. RyR2 is expressed in hair cells. Caffeine (5 mM) induces an increase in intracellular Ca^{2+} in SGNs in the absence of external Ca^{2+}. Application of Ryanodine (0.05 - 1 μM) may induce similar increases in $[Ca^{2+}]_i$. These data suggest RyR-mediated Ca^{2+} release from intracellular stores in SGNs [118, 143]. The finding that AMPA receptor agonist-induced increases in $[Ca^{2+}]_i$ in SGNs can be reduced by prior depletion of intracellular Ca^{2+} stores with caffeine, thapsigargin or ryanodine, indicates CICR mediated with RyRs [143].

Ca^{2+}-mediated cell death is usually acute and rapid [32, 61, 89, 194, 200, 217]. Disturbances of Ca^{2+} homeostasis in the cytoplasm, ER, or mitochondria can be harmful to cells [57]. Ca^{2+}-ATPase and the Na^+-Ca^{2+} exchanger on the plasma membrane remove cytosolic Ca^{2+} to the extracellular space [22, 34, 57]. $[Ca^{2+}]$ in the mitochondrial matrix is around 100 nM in resting cells. When cytosolic $[Ca^{2+}]$ rises, Ca^{2+} can enter mitochondria [60]. In mitochondria the Na^+-Ca^{2+} exchanger extrudes Ca^{2+} [57, 60, 84] whereas Na^+ entry may reserve the activity of the Na^+-Ca^{2+} exchanger [92] under pathological conditions [57]. Excessive Ca^{2+} entry may trigger changes in mitochondrial permeability. The sustained permeability changes may cause mitochondrial depolarization, inhibition of ATP production, and cell death [21, 45, 116]. Ca^{2+} in the nucleus is found to be rapidly equilibrated with cytosolic Ca^{2+} [133] and critically involved in regulations of the gene transcription and DNA metabolism [86, 135, 153]. Hair cells release glutamate to activate SGNs through both NMDA and kainate/AMPA-type glutamate receptors [63, 166]. Over-exposure to glutamate leads toxic effects on SGNs [47, 76, 157, 166]. Brief treatment with NMDA and kainate results in loss of IHC-SGN synapses and degeneration of axons of the distal type I SGNs. This result is also observed in SGNs following excitotoxic or noise stimulation in vitro [198].

Activated NMDARs are highly permeable to both Na^+ and Ca^{2+} [56, 132, 134]. We have observed that when extracellular solution contains more Na^+, NMDAR activation produces greater increases in both $[Na^+]_i$ and $[Ca^{2+}]_i$ [202]. Activation of NMDARs may increase $[Ca^{2+}]_i$ by 100 nM in neurons. Following pre-application of thapsigargin (0.1 μM), which depletes intracellular stores of Ca^{2+} by blocking Ca^{2+} re-uptake, $[Ca^{2+}]_i$ is increased by 60 nM after the NMDAR activation. The NMDAR activation only produces 35 nM increases in $[Ca^{2+}]_i$ in neurons when extracellular $[Na^+]$ is reduced to 10 mM [202]. NMDAR activations induce no Ca^{2+} increase in neurons if Ca^{2+}

influx is blocked by removal of extracellular Ca^{2+} [38, 196]. The increase in $[Ca^{2+}]_i$ induced by Ca^{2+} release from intracellular stores during NMDAR activation in neurons bathed with extracellular solution containing 10 mM Na^+ is significantly reduced when compared with that in neurons bathed with extracellular solution containing 145 mM Na^+ [202]. These data have suggested that CICR from intracellular stores during NMDAR activation may be regulated by Na^+ influx. Furthermore, it is found that intra-axonal Ca^{2+} release during ischemia in rat optic nerves mainly depends upon Na^+ influx, which may stimulate the mitochondrial Na^+/Ca^{2+} exchanger driven in the Na^+ import/Ca^{2+} export mode, positively modulate ryanodine receptors, and promote IP3 generation by phospholipase C [149]. Previous studies have shown that intracellular Na^+ acts as an up-regulator of NMDARs [75, 207, 210, 211] whereas increases of intracellular Ca^{2+} inhibits NMDAR activity [56, 132, 134]. Through Na^+ influx NMDARs can be up-regulated by AMPARs, voltage-gated Na^+ channels, non-selective cation channels and remote NMDARs [75, 207, 210, 211]. Moreover it is found that the up-regulation of NMDARs induced by activation of voltage-gated Na^+ channels depends on Na^+ influx and activation of Src family kinases (SFKs) [75, 207, 210, 211]. Although the role played by intracellular Na^+ in the regulation of SGN survival and functions remains in puzzle, a large amount of data obtained from investigations in other cells including various type neurons has demonstrated that significant increases in intracellular Na^+ is a characteristic event associated with cell injury [15-18, 70, 172, 183]. Therefore, we hypothesize that Na^+ influx and/or changes in $[Na^+]_i$ may also be critical events, which can be induced by changes in functions of voltage-gated Na^+ channels and cause the pathophysiological alterations associated with tinnitus and hearing loss. Thus, further detailed characterization of effects of Na^+ influx and/or changes in $[Na^+]_i$ in SGNs will not only lay an important basis for understanding mechanisms underlying the development of tinnitus and hearing loss but also for revealing new therapeutic targets to treat these clinical problems.

3. The Regulation of Voltage-Gated Na^+ and K^+ Currents by SFKs in SGNs

SFKs are protein tyrosine kinases. More and more data have shown that SFKs serve as a convergent point of multiple, diverse signaling pathways that regulate neuronal development and function [81, 94, 170, 193, 209]. Detailed studies have indicated that via multiple (direct or indirect) mechanisms, numerous SFK molecules are closely linked with and ready to act on their substrates, e.g., NMDARs [77, 81, 82, 104, 186, 211]. We have noted that the regulation of NMDARs by intracellular Na^+ is regulated by SFKs [207, 210, 211]. The inhibition of SFKs results in a reduction of the Na^+ effect on NMDARs. Conversely, the Na^+ effect is enhanced when SFK activity is up-regulated [207, 210, 211]. Application of SFK inhibitors has been found to be effective for preventing noise-induced hearing loss [25, 87]. While details of the mechanisms underlying the effect of SFK inhibitors remain to be clarified, seeking effective therapeutic approaches that target SFKs has been a major focus of pharmaceutical research [25, 39, 87, 125, 154, 170, 209].

SFKs contain nine members including Src, Fyn, Lck, Lyn, and Yes. These kinases are highly expressed in the nervous system and play important roles in the regulation of neural development and synaptic transmission [33, 81, 103, 170, 193, 209]. All the members of SFKs have a common topology: an N-terminus associated with the membrane, unique domain, SH2 and SH3 domains, kinase domain and a regulatory C-terminal region [31, 33, 103]. Diverse signaling pathways can regulate SFKs via regulating the function of each domain. Protein tyrosine phosphatase α (PTPα) is one of phosphatases which dephosphorylates phosphorylated C-terminal tyrosine [pY527 in chicken cellular Src (c-Src)] of SFKs, blocks the SH2 binding to the C-terminal and increases SFK activity [33, 81, 115, 155]. Conversely, endogenous C-terminal Src kinase (Csk) phosphorylates the C-terminal of SFKs, promotes the SH2 domain binding to the C-terminal and reduces SFK activity [33, 81, 151, 152, 204].

Earlier studies have shown that intra-domain interactions in SFKs (e.g., the SH2 domain binds to the phosphorylated C-terminus, or the SH2 kinase linker binds to the SH3 domain) may lock the kinases in a closed conformation, disrupt the kinase active site, and inactivate SFKs [12, 28, 31, 33, 40, 41, 43, 62]. Recently, it has been found that in active Fps kinase, the SH2 domain binds to the kinase N-terminal lobe, and positions the kinase αC helix in an active configuration [69]. In active cellular Abl (c-Abl) tyrosine kinase, the SH2 and SH3 domains are redistributed and promoting the catalytic activity of the enzyme [69, 145]. We have found that the up-regulation of NMDARs induced by expression of constitutively active neuronal Src (n-Src, an alternative isoform of Src kinase containing a 6-amino acid insert in the SH3 domain, and expressing only in neurons [33, 130, 188]) is significantly reduced by dysfunctions of the SH2 and/or SH3 domains of the protein.

Furthermore, our studies have shown that dysfunctions of SH2 and/or SH3 domains reduce the activity of n-Src but produce no change in the binding of n-Src to its substrate NMDAR proteins [81, 82]. All the findings dealing with the regulation of SFKs have strongly suggested that SFKs may act as "graded" kinases, which can exhibit different levels of enzyme activity coinciding with activation states [31, 33, 81, 82, 103, 129] and serve as a convergent point regulating cellular functions [81, 94, 170, 193, 209].

In different types of cells, e.g., vascular smooth muscle, HEK-293, tsA-201, renal, Schwann cells or cervical ganglion neurons (CGNs), SFKs show different effects on voltage-gated K^+ or Na^+ currents. This inconsistence may be due to different types of channels expressed [5, 6, 20, 101, 117, 121, 122, 178, 182, 189, 191]. Thus, the SFK regulation of voltage-gated K^+ and Na^+ currents appears to be "cell-type specific". The expression of voltage-gated Na^+ channels in rat SGNs exhibits a unique pattern [66, 73]. This may be related to the independent embryological origins of SGNs [66, 73]. Our immunofluoresence labeling of SFKs in cultured SGNs shows that SFK co-labeling can be found in all SGNs labeled with an antibody against neurofilament 200 [68].

We have recorded voltage-gated Na^+ and/or K^+ currents in cultured SGNs in the whole-cell configuration [68]. In order to determine how voltage-gated Na^+ or K^+ channels are regulated by endogenous SFKs in SGNs, effects produced by bath application of SFK inhibitors (PP2 [13, 64, 82, 106, 204] or SU6656 [26, 204]) or by intracellular delivery of a peptide which may inhibit Src activity (Src40-58 [103, 125, 169, 193]) or activate SFKs (SFK activator peptide [103, 123, 208]) have been investigated. For intracellular delivery of peptides, the peptides are added into the intracellular solution filling recording electrodes [68]. Effects of bath application of PP3 (the inactive form of PP2) or intracellular application of the scramble form peptide of Src40-58 or non-phosphorylated form of the SFK activator peptide are observed as control for PP2, Src40-58 or the SFK activator peptide [68].

Our data show that voltage-gated Na^+ currents are inhibited following the bath application of the SFK inhibitor, PP2 or SU6656 [68]. In contrast to this effect, voltage-gated K^+ currents in SGNs appear not affected with either PP2 or SU6656 application. SGNs possess all K^+ channels responsible for the depolarization-activated K^+ currents [2, 14, 99, 167]. It is known that depending upon the channel structure [74, 117] and intracellular signaling [182], some types of voltage-gate K^+ channels can be up-regulated [50, 121, 122, 160, 178] and some down-regulated [8, 35, 50, 83, 93, 184] by SFKs. However, detailed mechanisms underlying the effects of SFKs on voltage-gated K^+ currents in SGNs remain unclear. One possibility might be that the activity of endogenous SFKs regulating voltage-gate K^+ channels could be too low to be detected with application of SFK inhibitors. Another possibility might be that when SFKs were inhibited by application of PP2 or SU6656, some K^+ channels might be up- and some others down-regulated, whereby recorded voltage-gated K^+ currents in summation might maintain at the same level in SGNs. Consistently, it has been found that application of the wide spectrum protein tyrosine kinase inhibitor, genistein, does not alter the resting membrane potential but inhibits voltage-gated Na^+ currents in cervical ganglia neurons (CGNs) [101].

In studies of the regulation of voltage-gated Na^+ channels by SFKs, it is also found that different types of Na^+ channels may be differentially regulated by SFKs. Fyn, a member of SFKs, associates with Na(V)1.2 channels [6, 20]. Fyn over-expression causes phosphorylation of multiple tyrosine residues in Na(V)1.2 and Na(V)1.5 channels. The over-expression of Fyn inhibits the activity of Na(V)1.2 [6, 20] and Na(V)1.5 [5] channels co-expressed in tsA-201 and HEK-293 cells. Furthermore, it is found that the tyrosine phosphorylation of these channels promotes their voltage-dependent steady-state inactivation [5, 6, 20]. In contrast, the activity of the Na(V)1.1 channel is not affected by Fyn [20].

The finding that application of genistein inhibits voltage-gated Na^+ currents in CGNs [101] (where Na(V) 1.7 channels are highly expressed [55, 101, 190]) has suggested that endogenous protein tyrosine kinases may up-regulate basal voltage-gated Na^+ currents. Our data in SGNs has shown that voltage-gated Na^+ current amplitudes are reduced by application of SFK inhibitors, but increased by application of a SFK activator [68]. The finding that no significant effect on inward current can be induced by the SFK activator peptide applied into neurons treated with TTX, implies that TTX-sensitive voltage-gated Na^+ channels in SGNs may be tonically up-regulated by endogenous SFKs [68].

The data reported from more detailed studies regarding the regulation of the voltage-dependent activation and steady-state inactivation of Na^+ channels by SFKs, seem not consistent. For example, over-expression of Fyn (which causes phosphorylation of the tyrosine residues 1497 and 1498 in the loop regulating the steady-state inactivation of Na^+ channels) is found to promote the inactivation without affecting the voltage-dependent activation of Na(V) 1.2 or Na(V)1.5 channels co-expressed in HEK-293 cells [5, 20]. However, application of genestein is found to cause a right shift of the voltage-dependent activation of Na^+ channels in CGNs [101]. In SGNs, our data show that under basal conditions the application of SFK inhibitors does not affect voltage-dependent activation but enhances the steady-state inactivation, and delays the recovery of Na^+ currents [68]. Conversely, up-regulating SFKs by application of a SFK activator promotes the voltage-dependent activation but does not affect the steady-state inactivation or recovery of the Na^+ currents [68]. Since the intracellular SFK activator-induced change in the voltage-dependent activation can be abolished by bath co-application of a SFK inhibitor, we conclude that all the voltage-dependent activation, steady-state inactivation and recovery of Na^+ currents in SGNs are regulated by endogenous SFKs. The treatment potentiating the activity of SFKs enhances the voltage-dependent activation while the treatment inhibiting SFKs reduces the steady-state inactivation and recovery of Na^+ currents. These findings imply a possibility that the steady-state inactivation and the recovery of Na^+ channels may be controlled by activated SFKs. SFKs involved in the regulation of voltage-dependent activation of Na^+ channels may be less active or inactive. This hypothesis has been supported by findings that endogenous SFKs are regulated by a variety of mechanisms and therefore may exhibit different levels of enzyme activity coinciding with activation states [31, 33, 82, 103, 129]. Thus, SFKs can exhibit a bifunctional regulation of Na^+, but not K^+, currents in SGNs.

Encoding sound signals into a neuronal code is a critical process of hearing. Voltage-gated Na^+ channels play a key role in the generation and propagation of action potential. Findings that the voltage-dependent activation, steady-state inactivation and recovery of Na^+ currents in SGNs are regulated by endogenous SFKs have strongly suggested that SFKs are involved in regulation of the excitability and firing behavior (such as discharge frequency and pattern in response to sound stimuli) of SGNs. Thus, further characterizing mechanisms underlying the regulation of voltage-gated Na^+ and K^+ channels by SFKs in SGNs will be required for advancing the understanding of pathological processes related to tinnitus and/or hearing loss.

4. Potential Roles of SFKs in Tinnitus and Hearing Loss

The survival of SGNs critically depends on intact hair cells. Following hair cell loss, SGNs undergo apoptosis [7, 110, 114, 171]. The cochlear implant, which compensates for lost hair cell function by directly stimulating the auditory nerve, is found to be an effective way to treat many forms of deafness [11]. Application of neurotrophins such as NT-3 and BDNF [71, 138, 144, 161, 181] or direct electrical stimulation to SGNs [88, 112, 113, 126, 139, 175] may also significantly reduce SGN death induced by hair cell loss. It is found that delayed loss of SGNs following acoustic trauma [108] may be a result of excitotoxic damage to SGNs contacting synapses with IHCs [109]. Brief treatment with NMDA/kainite causes loss of IHC-SGN synapses and damage of SGNs [198]. This effect can be diminished by application of BDNF or NT-3, suggesting a therapeutic intervention for acoustic trauma [198].

Endogenous NT-3 and BDNF have been found to support SGN survival during development [80, 159]. These neurotrophins remain expressed in the postnatal cochlea. Neurotrophin null mutants show specific patterns of neuronal loss along the cochlea and reduced radial fiber growth [206]. Through the activation of trkB and trkC receptors, NT-3 and BDNF in SGNs are involved in the regulation of the expression and functions of ion channels and hair cell-SGN synapses [1, 80, 96, 159, 198]. Although differences in effects of NT-3 versus BDNF have been found on SGNs, recent studies have strongly suggested that these neurotrophins play important roles in the regulation of auditory function [1, 80, 111, 159, 195]. These neurotrophins are necessary for the establishment of characteristic firing features of SGNs [1]. Application of BDNF or NT-3 can not only prevent SGN degeneration following hair cell loss [159] but also promote SGN survival, axon regrowth, and function in deafened animals [111, 159, 195]. Endogenous NT-3 promotes the regeneration of synapses between IHCs and type I SGNs after excitotoxic trauma in vitro [198]. Great insights for developing therapeutic approaches targeting the regulation of SGN functions by these neurotrophins have been revealed [1, 80, 111, 159, 195].

More detailed mechanistic investigations in other types of cells/neurons have demonstrated that SFKs such as Src and Fyn play critical roles in trkB and trkC signaling pathways [6, 20, 97, 98, 102, 140, 205, 215]. Through Src signaling the growth factor receptor tyrosine kinases acutely regulate Na^+ channels in neurons [91]. SFKs can be activated by TrkB and, in turn, SFKs promote TrkB activation [97]. Depending upon the autophosphorylation of the intracellular domain of TrkB, the SH2 domain of Fyn binds to TrkB [98]. Fyn also plays an important role in the interaction between TrkB and the NMDAR NR2B subunit. BDNF-TrkB signaling is critically involved in the process of learning and memory [140, 205]. Via the channel-associated Fyn BDNF regulates Nav1.2 channels [6, 20]. Through the TrkB-Src/PLC-γ1 pathway, BDNF regulates the release of glutamate from cortical neurons [215]. c-Src is required for TrkC-induced activation of the phosphatidylinositol 3-kinase (PI3K)-AKT pathway [102]. Although it still needs direct evidence to demonstrate whether and how SFKs are involved in the regulation of auditory functions by neurotrophins in SGNs, based on above mentioned data it would be reasonable to hypothesize that SFKs may be critically involved in the pathophysiological changes in auditory functions.

Conclusion and Future Studies

Local conditioning factors such as the release of BDNF and/or NT-3, are important for regulation of expressions of various voltage- and ligand-gated ionic channels in SGNs [127, 128]. The inhibition of SFKs may reduce voltage-gated Na^+ current amplitude, enhance steady-state inactivation and delay the recovery of Na^+ channels. Conversely, the augmentation of SFK activity may potentiate Na^+ current and promote the voltage-dependent activation [68]. Thus, the excitability and firing behavior (including discharge frequency and pattern in response to sound stimuli) of SGNs may be regulated by SFKs. Although the underlying mechanism remains unknown, the finding that the inhibition of SFKs may prevent noise-induced hearing loss [25, 87] has provided another line of evidence supporting this possibility. Moreover, it has been reported that the up-regulation of NMDARs induced by activation of voltage-gated Na^+ channels depends on both Na^+ influx and the activity of SFKs [75, 207, 210, 211]. SFKs act as critical molecules involved in multiple signaling pathways controlling neuronal life and functions, which include the regulations of actions of intracellular Na^+ and Na^+ influx through either ligand or voltage-gated Na^+ channels. Thus, further characterizing roles of SFKs and Na^+ influx in the regulation of auditory functions in disease status will be needed for understanding the development of tinnitus and/or hearing loss.

References

[1] Adamson, C. L., Reid, M. A., Davis, R. L. (2002) Opposite actions of brain-derived neurotrophic factor and neurotrophin-3 on firing features and ion channel composition of murine spiral ganglion neurons. *J. Neurosci.* 22: 1385-1396.

[2] Adamson, C. L., Reid, M. A., Mo, Z. L., Bowne-English, J., Davis, R. L. (2002) Firing features and potassium channel content of murine spiral ganglion neurons vary with cochlear location. *J. Comp. Neurol.* 447: 331-350.

[3] Agrawal, S. K., Fehlings, M. G. (1996) Mechanisms of secondary injury to spinal cord axons in vitro: role of Na+, Na(+)-K(+)-ATPase, the Na(+)-H+ exchanger, and the Na(+)-Ca2+ exchanger. *J. Neurosci.* 16: 545-552.

[4] Agrawal, S. K., Fehlings, M. G. (1997) The effect of the sodium channel blocker QX-314 on recovery after acute spinal cord injury. *J. Neurotrauma.* 14: 81-88.

[5] Ahern, C. A., Zhang, J. F., Wookalis, M. J., Horn, R. (2005) Modulation of the cardiac sodium channel NaV1.5 by Fyn, a Src family tyrosine kinase. *Circ. Res.* 96: 991-998.

[6] Ahn, M., Beacham, D., Westenbroek, R. E., Scheuer, T., Catterall, W. A. (2007) Regulation of Na(v)1.2 channels by brain-derived neurotrophic factor, TrkB, and associated Fyn kinase. *J. Neurosci.* 27: 11533-11542.

[7] Alam, S. A., Robinson, B. K., Huang, J., Green, S. H. (2007) Prosurvival and proapoptotic intracellular signaling in rat spiral ganglion neurons in vivo after the loss of hair cells. *J. Comp. Neurol.* 503: 832-852.

[8] Alioua, A., Mahajan, A., Nishimaru, K., Zarei, M. M., Stefani, E., Toro, L. (2002) Coupling of c-Src to large conductance voltage- and Ca2+-activated K+ channels as a new mechanism of agonist-induced vasoconstriction. *Proc. Natl. Acad. Sci. U S A* 99: 14560-14565.

[9] Altschuler, R. A., Sheridan, C. E., Horn, J. W., Wenthold, R. J. (1989) Immunocytochemical localization of glutamate immunoreactivity in the guinea pig cochlea. *Hear. Res.* 42: 167-173.

[10] Appelgren, L., Janson, M., Nitescu, P., Curelaru, I. (1996) Continuous intracisternal and high cervical intrathecal bupivacaine analgesia in refractory head and neck pain [see comments]. *Anesthesiology* 84: 256-272.

[11] Appler, J. M., Goodrich, L. V. (2011) Connecting the ear to the brain: Molecular mechanisms of auditory circuit assembly. *Prog. Neurobiol.* 93: 488-508.

[12] Ayrapetov, M. K., Wang, Y. H., Lin, X., Gu, X., Parang, K., Sun, G. (2006) Conformational basis for SH2-Tyr(P)527 binding in Src inactivation. *J. Biol. Chem.* 281: 23776-23784.

[13] Bain, J., McLauchlan, H., Elliott, M., Cohen, P. (2003) The specificities of protein kinase inhibitors: an update. *Biochem. J.* 371: 199-204.

[14] Bakondi, G., Por, A., Kovacs, I., Szucs, G., Rusznak, Z. (2008) Voltage-gated K+ channel (Kv) subunit expression of the guinea pig spiral ganglion cells studied in a newly developed cochlear free-floating preparation. *Brain Res.* 1210: 148-162.

[15] Ballard-Croft, C., Carlson, D., Maass, D. L., Horton, J. W. (2004) Burn trauma alters calcium transporter protein expression in the heart. *J. Appl. Physiol.* 97: 1470-1476.

[16] Banasiak, K. J., Burenkova, O., Haddad, G. G. (2004) Activation of voltage-sensitive sodium channels during oxygen deprivation leads to apoptotic neuronal death. *Neuroscience* 126: 31-44.

[17] Baptiste, D. C., Fehlings, M. G. (2007) Update on the treatment of spinal cord injury. *Prog. Brain Res.* 161: 217-233.

[18] Bauer, R., Walter, B., Fritz, H., Zwiener, U. (1999) Ontogenetic aspects of traumatic brain edema--facts and suggestions. *Exp. Toxicol. Pathol.* 51: 143-150.

[19] Baxter, K. A., Church, J. (1996) Characterization of acid extrusion mechanisms in cultured fetal rat hippocampal neurones. *J. Physiol. (Lond)* 493: 457-470.

[20] Beacham, D., Ahn, M., Catterall, W. A., Scheuer, T. (2007) Sites and molecular mechanisms of modulation of Na(v)1.2 channels by Fyn tyrosine kinase. *J. Neurosci.* 27: 11543-11551.

[21] Bernardi, P., Rasola, A. (2007) Calcium and cell death: the mitochondrial connection. *Subcell. Biochem.* 45: 481-506.

[22] Berridge, M. J., Bootman, M. D., Roderick, H. L. (2003) Calcium signalling: dynamics, homeostasis and remodelling. *Nat. Rev. Mol. Cell Biol* 4: 517-529.

[23] Bhattacharjee, A., Joiner, W. J., Wu, M., Yang, Y., Sigworth, F. J., Kaczmarek, L. K. (2003) Slick (Slo2.1), a rapidly-gating sodium-activated potassium channel inhibited by ATP. *J. Neurosci.* 23: 11681-11691.

[24] Bhattacharjee, A., Kaczmarek, L. K. (2005) For K(+) channels, Na(+) is the new Ca(2+). *Trends Neurosci.* 28: 422-428.

[25] Bielefeld, E. C., Hynes, S., Pryznosch, D., Liu, J., Coleman, J. K., Henderson, D. (2005) A comparison of the protective effects of systemic administration of a pro-glutathione drug and a Src-PTK inhibitor against noise-induced hearing loss. *Noise. Health* 7: 24-30.

[26] Blake, R. A., Broome, M. A., Liu, X., Wu, J., Gishizky, M., Sun, L., Courtneidge, S. A. (2000) SU6656, a selective src family kinase inhibitor, used to probe growth factor signaling. *Mol. Cell Biol.* 20: 9018-9027.

[27] Blaustein, M. P., Fontana, G., Rogowski, R. S. (1996) The Na(+)-Ca2+ exchanger in rat brain synaptosomes. Kinetics and regulation. *Ann. N.Y. Acad. Sci.* 779:300-17: 300-317.

[28] Boggon, T. J., Eck, M. J. (2004) Structure and regulation of Src family kinases. *Oncogene* 23: 7918-7927.

[29] Boonstra, J., Moolenaar, W. H., Harrison, P. H., Moed, P., van der Saag, P. T., de Laat, S. W. (1983) Ionic responses and growth stimulation induced by nerve growth factor and epidermal growth factor in rat pheochromocytoma (PC12) cells. *J. Cell Biol.* 97: 92-98.

[30] Bortner, C. D., Cidlowski, J. A. (2003) Uncoupling cell shrinkage from apoptosis reveals that Na+ influx is required for volume loss during programmed cell death. *J. Biol. Chem.* 278: 39176-39184.

[31] Bradshaw, J. M. (2010) The Src, Syk, and Tec family kinases: distinct types of molecular switches. *Cell Signal.* 22: 1175-1184.

[32] Bredesen, D. E. (2000) Apoptosis: overview and signal transduction pathways. *J. Neurotrauma* 17: 801-810.

[33] Brown, M. T., Cooper, J. A. (1996) Regulation, substrates and functions of src. *Biochim. Biophys. Acta* 1287: 121-149.

[34] Carafoli, E., Santella, L., Branca, D., Brini, M. (2001) Generation, control, and processing of cellular calcium signals. *Crit. Rev. Biochem. Mol. Biol.* 36: 107-260.

[35] Cayabyab, F. S., Khanna, R., Jones, O. T., Schlichter, L. C. (2000) Suppression of the rat microglia Kv1.3 current by src-family tyrosine kinases and oxygen/glucose deprivation. *Eur. J. Neurosci.* 12: 1949-1960.

[36] Choi, D. W. (1993) NMDA receptors and AMPA/kainate receptors mediate parallel injury in cerebral cortical cultures subjected to oxygen-glucose deprivation. *Prog. Brain Res.* 96: 137-143.

[37] Choi, D. W. (1995) Calcium: still center-stage in hypoxic-ischemic neuronal death. *Trends. Neurosci.* 18: 58-60.

[38] Clapham, D. E. (1995) Calcium signaling. *Cell* 80: 259-268.

[39] Cohen, P. (2002) Protein kinases--the major drug targets of the twenty-first century? *Nat. Rev. Drug Discov.* 1: 309-315.

[40] Cole, P. A., Shen, K., Qiao, Y., Wang, D. (2003) Protein tyrosine kinases Src and Csk: a tail's tale. *Curr. Opin. Chem. Biol.* 7: 580-585.

[41] Cooper, J. A., Gould, K. L., Cartwright, C. A., Hunter, T. (1986) Tyr527 is phosphorylated in pp60c-src: implications for regulation. *Science* 231: 1431-1434.

[42] Couloigner, V., Fay, M., Djelidi, S., Farman, N., Escoubet, B., Runembert, I., Sterkers, O., Friedlander, G., Ferrary, E. (2001) Location and function of the epithelial Na channel in the cochlea. *Am. J. Physiol. Renal. Physiol.* 280: F214-F222.

[43] Cowan-Jacob, S. W., Fendrich, G., Manley, P. W., Jahnke, W., Fabbro, D., Liebetanz, J., Meyer, T. (2005) The crystal structure of a c-Src complex in an active conformation suggests possible steps in c-Src activation. *Structure.* 13: 861-871.

[44] Cox, J. J., Reimann, F., Nicholas, A. K., Thornton, G., Roberts, E., Springell, K., Karbani, G., Jafri, H., Mannan, J., Raashid, Y., Al Gazali, L., Hamamy, H., Valente, E. M., Gorman, S., Williams, R., McHale, D. P., Wood, J. N., Gribble, F. M., Woods, C. G. (2006) An SCN9A channelopathy causes congenital inability to experience pain. *Nature* 444: 894-898.

[45] Crompton, M. (1999) The mitochondrial permeability transition pore and its role in cell death. *Biochem. J.* 341 (Pt 2): 233-249.

[46] Cummins, T. R., Waxman, S. G. (1997) Downregulation of tetrodotoxin-resistant sodium currents and upregulation of a rapidly repriming tetrodotoxin-sensitive sodium current in small spinal sensory neurons after nerve injury. *J. Neurosci.* 17: 3503-3514.

[47] d'Aldin, C. G., Ruel, J., Assie, R., Pujol, R., Puel, J. L. (1997) Implication of NMDA type glutamate receptors in neural regeneration and neoformation of synapses after excitotoxic injury in the guinea pig cochlea. *Int. J. Dev. Neurosci.* 15: 619-629.

[48] Dallos, P., Wu, X., Cheatham, M. A., Gao, J., Zheng, J., Anderson, C. T., Jia, S., Wang, X., Cheng, W. H., Sengupta, S., He, D. Z., Zuo, J. (2008) Prestin-based outer hair cell motility is necessary for mammalian cochlear amplification. *Neuron* 58: 333-339.

[49] Darlington, C. L., Smith, P. F. (2007) Drug treatments for tinnitus. *Prog. Brain Res* 166: 249-262.

[50] Davis, M. J., Wu, X., Nurkiewicz, T. R., Kawasaki, J., Gui, P., Hill, M. A., Wilson, E. (2001) Regulation of ion channels by protein tyrosine phosphorylation. *Am. J. Physiol. Heart Circ. Physiol.* 281: H1835-H1862.

[51] Davis, R. L., Liu, Q. (2011) Complex primary afferents: What the distribution of electrophysiologically-relevant phenotypes within the

spiral ganglion tells us about peripheral neural coding. *Hear. Res.* 276: 34-43.

[52] Defourny, J., Lallemend, F., Malgrange, B. (2011) Structure and development of cochlear afferent innervation in mammals. *Am. J. Physiol. Cell Physiol.* 301: C750-C761

[53] Deng, L., Ding, D., Su, J., Manohar, S., Salvi, R. (2013) Salicylate Selectively Kills Cochlear Spiral Ganglion Neurons by Paradoxically Up-regulating Superoxide. *Neurotox. Res.*

[54] Dib-Hajj, S., Black, J. A., Cummins, T. R., Waxman, S. G. (2002) NaN/Nav1.9: a sodium channel with unique properties. *Trends Neurosci.* 25: 253-259.

[55] Dib-Hajj, S. D., Black, J. A., Waxman, S. G. (2009) Voltage-gated sodium channels: therapeutic targets for pain. *Pain Med.* 10: 1260-1269.

[56] Dingledine, R., Borges, K., Bowie, D., Traynelis, S. F. (1999) The glutamate receptor ion channels. *Pharmacol. Rev.* 51: 7-61.

[57] Dong, Z., Saikumar, P., Weinberg, J. M., Venkatachalam, M. A. (2006) Calcium in cell injury and death. *Annu. Rev. Pathol.* 1: 405-434.

[58] Drescher, M. J., Drescher, D. G. (1992) Glutamate, of the endogenous primary alpha-amino acids, is specifically released from hair cells by elevated extracellular potassium. *J. Neurochem.* 59: 93-98.

[59] Dryer, S. E. (2003) Molecular identification of the Na+-activated K+ channel. *Neuron* 37: 727-728.

[60] Duchen, M. R. (2000) Mitochondria and calcium: from cell signalling to cell death. *J. Physiol.* 529 Pt 1: 57-68.

[61] Duchen, M. R. (2012) Mitochondria, calcium-dependent neuronal death and neurodegenerative disease. *Pflugers Arch.* 464: 111-121.

[62] Eck, M. J., Atwell, S. K., Shoelson, S. E., Harrison, S. C. (1994) Structure of the regulatory domains of the Src-family tyrosine kinase Lck. *Nature* 368: 764-769.

[63] Eybalin, M. (1993) Neurotransmitters and neuromodulators of the mammalian cochlea. *Physiol. Rev.* 73: 309-373.

[64] Fang, X. Q., Xu, J., Feng, S., Groveman, B. R., Lin, S. X., Yu, X. M. (2011) The NMDA receptor NR1 subunit is critically involved in the regulation of NMDA receptor activity by C-terminal Src kinase (Csk). *Neurochem. Res.* 36: 319-326.

[65] Fehlings, M. G., Agrawal, S. (1995) Role of sodium in the pathophysiology of secondary spinal cord injury. *Spine.* 20: 2187-2191.

[66] Fekete, D. M., Wu, D. K. (2002) Revisiting cell fate specification in the inner ear. *Curr. Opin. Neurobiol.* 12: 35-42.

[67] Feng, H., Yin, S. H., Tang, A. Z., Tan, S. H. (2011) Salicylate initiates apoptosis in the spiral ganglion neuron of guinea pig cochlea by activating caspase-3. *Neurochem. Res.* 36: 1108-1115.

[68] Feng, S., Pflueger, M., Lin, S. X., Groveman, B. R., Su, J., Yu, X. M. (2012) Regulation of voltage-gated sodium current by endogenous Src family kinases in cochlear spiral ganglion neurons in culture. *Pflugers Arc.h* 463: 571-584.

[69] Filippakopoulos, P., Kofler, M., Hantschel, O., Gish, G. D., Grebien, F., Salah, E., Neudecker, P., Kay, L. E., Turk, B. E., Superti-Furga, G., Pawson, T., Knapp, S. (2008) Structural coupling of SH2-kinase domains links Fes and Abl substrate recognition and kinase activation. *Cell* 134: 793-803.

[70] Friedman, J. E., Haddad, G. G. (1994) Anoxia induces an increase in intracellular sodium in rat central neurons in vitro. *Brain Res.* 663: 329-334.

[71] Fritzsch, B., Silos-Santiago, I., Bianchi, L. M., Farinas, I. (1997) The role of neurotrophic factors in regulating the development of inner ear innervation. *Trends Neurosci.* 20: 159-164.

[72] Fryatt, A. G., Mulheran, M., Egerton, J., Gunthorpe, M. J., Grubb, B. D. (2011) Ototrauma induces sodium channel plasticity in auditory afferent neurons. *Mol. Cell Neurosci.* 48: 51-61.

[73] Fryatt, A. G., Vial, C., Mulheran, M., Gunthorpe, M. J., Grubb, B. D. (2009) Voltage-gated sodium channel expression in rat spiral ganglion neurons. *Mol. Cell Neurosci.* 42: 399-407.

[74] Gamper, N., Stockand, J. D., Shapiro, M. S. (2003) Subunit-specific modulation of KCNQ potassium channels by Src tyrosine kinase. *J. Neurosci.* 23: 84-95.

[75] George, J., Dravid, S. M., Prakash, A., Xie, J., Peterson, J., Jabba, S. V., Baden, D. G., Murray, T. F. (2009) Sodium channel activation augments NMDA receptor function and promotes neurite outgrowth in immature cerebrocortical neurons. *J. Neurosci.* 29: 3288-3301.

[76] Gil-Loyzaga, P., Pujol, R. (1990) Neurotoxicity of kainic acid in the rat cochlea during early developmental stages. *Eur. Arch. Otorhinolaryngol.* 248: 40-48.

[77] Gingrich, J. R., Pelkey, K. A., Fam, S. R., Huang, Y., Petralia, R. S., Wenthold, R. J., Salter, M. W. (2004) Unique domain anchoring of Src to synaptic NMDA receptors via the mitochondrial protein NADH dehydrogenase subunit 2. *Proc. Natl. Acad. Sci. U S A* 101: 6237-6242.

[78] Glowatzki, E., Fuchs, P. A. (2002) Transmitter release at the hair cell ribbon synapse. *Nat. Neurosci.* 5: 147-154.

[79] Godfrey, D. A., Carter, J. A., Berger, S. J., Matschinsky, F. M. (1976) Levels of putative transmitter amino acids in the guinea pig cochlea. *J. Histochem. Cytochem.* 24: 468-470.

[80] Green, S. H., Bailey, E., Wang, Q., Davis, R. L. (2012) The Trk A, B, C's of neurotrophins in the cochlea. *Anat. Rec. (Hoboken.)* 295: 1877-1895.

[81] Groveman, B. R., Feng, S., Fang, X. Q., Pflueger, M., Lin, S. X., Bienkiewicz, E. A., Yu, X. (2012) The regulation of N-methyl-D-aspartate receptors by Src kinase. *FEBS J.* 279: 20-28.

[82] Groveman, B. R., Xue, S., Marin, V., Xu, J., Ali, M. K., Bienkiewicz, E. A., Yu, X. M. (2011) Roles of the SH2 and SH3 domains in the regulation of neuronal Src kinase functions. *FEBS J.* 278: 643-653.

[83] Gu, R. M., Wei, Y., Falck, J. R., Krishna, U. M., Wang, W. H. (2001) Effects of protein tyrosine kinase and protein tyrosine phosphatase on apical K(+) channels in the TAL. *Am. J. Physiol. Cell Physiol.* 281: C1188-C1195.

[84] Gunter, T. E., Yule, D. I., Gunter, K. K., Eliseev, R. A., Salter, J. D. (2004) Calcium and mitochondria. *FEBS Lett.* 567: 96-102.

[85] Hains, B. C., Saab, C. Y., Lo, A. C., Waxman, S. G. (2004) Sodium channel blockade with phenytoin protects spinal cord axons, enhances axonal conduction, and improves functional motor recovery after contusion SCI. *Exp. Neurol.* 188: 365-377.

[86] Hardingham, G. E., Cruzalegui, F. H., Chawla, S., Bading, H. (1998) Mechanisms controlling gene expression by nuclear calcium signals. *Cell Calcium* 23: 131-134.

[87] Harris, K. C., Hu, B., Hangauer, D., Henderson, D. (2005) Prevention of noise-induced hearing loss with Src-PTK inhibitors. *Hear. Res.* 208: 14-25.

[88] Hartshorn, D. O., Miller, J. M., Altschuler, R. A. (1991) Protective effect of electrical stimulation in the deafened guinea pig cochlea. *Otolaryngol. Head Neck Surg.* 104: 311-319.

[89] Hegarty, J. L., Kay, A. R., Green, S. H. (1997) Trophic support of cultured spiral ganglion neurons by depolarization exceeds and is additive with that by neurotrophins or cAMP and requires elevation of [Ca2+]i within a set range. *J. Neurosci.* 17: 1959-1970.

[90] Henderson, D., Hamernik, R. P. (1995) Biologic bases of noise-induced hearing loss. *Occup. Med.* 10: 513-534.

[91] Hilborn, M. D., Vaillancourt, R. R., Rane, S. G. (1998) Growth factor receptor tyrosine kinases acutely regulate neuronal sodium channels through the src signaling pathway. *J. Neurosci.* 18: 590-600.

[92] Hilgemann, D. W., Collins, A., Matsuoka, S. (1992) Steady-state and dynamic properties of cardiac sodium-calcium exchange. Secondary modulation by cytoplasmic calcium and ATP. *J. Gen. Physiol.* 100: 933-961.

[93] Holmes, T. C., Fadool, D. A., Ren, R., Levitan, I. B. (1996) Association of Src tyrosine kinase with a human potassium channel mediated by SH3 domain. *Science* 274: 2089-2091.

[94] Hossain, M. I., Kamaruddin, M. A., Cheng, H. C. (2012) Aberrant regulation and function of Src family tyrosine kinases: their potential contributions to glutamate-induced neurotoxicity. *Clin. Exp. Pharmacol. Physiol.* 39: 684-691.

[95] Hossain, W. A., Antic, S. D., Yang, Y., Rasband, M. N., Morest, D. K. (2005) Where is the spike generator of the cochlear nerve? Voltage-gated sodium channels in the mouse cochlea. *J. Neurosci.* 25: 6857-6868.

[96] Huang, E. J., Reichardt, L. F. (2003) Trk receptors: roles in neuronal signal transduction. *Annu. Rev. Biochem.* 72: 609-642.

[97] Huang, Y. Z., McNamara, J. O. (2010) Mutual regulation of Src family kinases and the neurotrophin receptor TrkB. *J. Biol. Chem.* 285: 8207-8217.

[98] Iwasaki, Y., Gay, B., Wada, K., Koizumi, S. (1998) Association of the Src family tyrosine kinase Fyn with TrkB. *J. Neurochem.* 71: 106-111.

[99] Jagger, D. J., Housley, G. D. (2002) A-type potassium currents dominate repolarisation of neonatal rat primary auditory neurones in situ. *Neuroscience* 109: 169-182.

[100] Jagger, D. J., Housley, G. D. (2003) Membrane properties of type II spiral ganglion neurones identified in a neonatal rat cochlear slice. *J. Physiol.* 552: 525-533.

[101] Jia, Z., Jia, Y., Liu, B., Zhao, Z., Jia, Q., Liang, H., Zhang, H. (2008) Genistein inhibits voltage-gated sodium currents in SCG neurons through protein tyrosine kinase-dependent and kinase-independent mechanisms. *Pflugers. Arch.* 456: 857-866.

[102] Jin, W., Yun, C., Jeong, J., Park, Y., Lee, H. D., Kim, S. J. (2008) c-Src is required for tropomyosin receptor kinase C (TrkC)-induced activation of the phosphatidylinositol 3-kinase (PI3K)-AKT pathway. *J. Biol. Chem.* 283: 1391-1400.

[103] Kalia, L. V., Gingrich, J. R., Salter, M. W. (2004) Src in synaptic transmission and plasticity. *Oncogene* 23: 8007-8016.

[104] Kalia, L. V., Salter, M. W. (2003) Interactions between Src family protein tyrosine kinases and PSD-95. *Neuropharmacology* 45: 720-728.

[105] Kandler, K., Clause, A., Noh, J. (2009) Tonotopic reorganization of developing auditory brainstem circuits. *Nat. Neurosci.* 12: 711-717.

[106] Karni, R., Mizrachi, S., Reiss-Sklan, E., Gazit, A., Livnah, O., Levitzki, A. (2003) The pp60c-Src inhibitor PP1 is non-competitive against ATP. *FEBS Lett.* 537: 47-52.

[107] Koch, R. A., Barish, M. E. (1994) Perturbation of intracellular calcium and hydrogen ion regulation in cultured mouse hippocampal neurons by reduction of the sodium ion concentration gradient. *J. Neurosci.* 14: 2585-2593.

[108] Kujawa, S. G., Liberman, M. C. (2006) Acceleration of age-related hearing loss by early noise exposure: evidence of a misspent youth. *J. Neurosci.* 26: 2115-2123.

[109] Kujawa, S. G., Liberman, M. C. (2009) Adding insult to injury: cochlear nerve degeneration after "temporary" noise-induced hearing loss. *J. Neurosci.* 29: 14077-14085.

[110] Ladrech, S., Guitton, M., Saido, T., Lenoir, M. (2004) Calpain activity in the amikacin-damaged rat cochlea. *J. Comp. Neurol.* 477: 149-160.

[111] Leake, P. A., Hradek, G. T., Hetherington, A. M., Stakhovskaya, O. (2011) Brain-derived neurotrophic factor promotes cochlear spiral ganglion cell survival and function in deafened, developing cats. *J. Comp. Neurol.* 519: 1526-1545.

[112] Leake, P. A., Hradek, G. T., Rebscher, S. J., Snyder, R. L. (1991) Chronic intracochlear electrical stimulation induces selective survival of spiral ganglion neurons in neonatally deafened cats. *Hear. Res* 54: 251-271.

[113] Leake, P. A., Hradek, G. T., Snyder, R. L. (1999) Chronic electrical stimulation by a cochlear implant promotes survival of spiral ganglion neurons after neonatal deafness. *J. Comp. Neurol.* 412: 543-562.

[114] Lee, J. E., Nakagawa, T., Kim, T. S., Iguchi, F., Endo, T., Dong, Y., Yuki, K., Naito, Y., Lee, S. H., Ito, J. (2003) A novel model for rapid induction of apoptosis in spiral ganglions of mice. *Laryngoscope* 113: 994-999.

[115] Lei, G., Xue, S., Chery, N., Liu, Q., Xu, J., Kwan, C. L., Fu, Y., Lu, Y. M., Liu, M., Harder, K. H., Yu, X. M. (2002) Gain control of N-methyl-

D-aspartate receptor activity by receptor-like protein tyrosine phopshatase alpha. *The EMBO J.* 21: 2977-2989.

[116] Lemasters, J. J., Theruvath, T. P., Zhong, Z., Nieminen, A. L. (2009) Mitochondrial calcium and the permeability transition in cell death. *Biochim. Biophys. Acta* 1787: 1395-1401.

[117] Li, Y., Langlais, P., Gamper, N., Liu, F., Shapiro, M. S. (2004) Dual phosphorylations underlie modulation of unitary KCNQ K(+) channels by Src tyrosine kinase. *J. Biol. Chem.* 279: 45399-45407.

[118] Liang, Y., Huang, L., Yang, J. (2009) Differential expression of ryanodine receptor in the developing rat cochlea. *Eur. J. Histochem.* 53: e30.

[119] Lin, X. (1997) Action potentials and underlying voltage-dependent currents studied in cultured spiral ganglion neurons of the postnatal gerbil. *Hear. Res.* 108: 157-179.

[120] Lin, X., Chen, S., Tee, D. (1998) Effects of quinine on the excitability and voltage-dependent currents of isolated spiral ganglion neurons in culture. *J. Neurophysiol.* 79: 2503-2512.

[121] Ling, S., Sheng, J. Z., Braun, A. P. (2004) The calcium-dependent activity of large-conductance, calcium-activated K+ channels is enhanced by Pyk2- and Hck-induced tyrosine phosphorylation. *Am. J. Physiol. Cell Physiol.* 287: C698-C706.

[122] Ling, S., Woronuk, G., Sy, L., Lev, S., Braun, A. P. (2000) Enhanced activity of a large conductance, calcium-sensitive K+ channel in the presence of Src tyrosine kinase. *J. Biol. Chem.* 275: 30683-30689.

[123] Liu, X., Brodeur, S. R., Gish, G., Songyang, Z., Cantley, L. C., Laudano, A. P., Pawson, T. (1993) Regulation of c-Src tyrosine kinase activity by the Src SH2 domain. *Oncogene* 8: 1119-1126.

[124] Liu, X., Stan, L. L. (2004) Sodium-activated potassium conductance participates in the depolarizing afterpotential following a single action potential in rat hippocampal CA1 pyramidal cells. *Brain Res.* 1023: 185-192.

[125] Liu, X. J., Gingrich, J. R., Vargas-Caballero, M., Dong, Y. N., Sengar, A., Beggs, S., Wang, S. H., Ding, H. K., Frankland, P. W., Salter, M. W. (2008) Treatment of inflammatory and neuropathic pain by uncoupling Src from the NMDA receptor complex. *Nat. Med.* 14: 1325-1332.

[126] Lousteau, R. J. (1987) Increased spiral ganglion cell survival in electrically stimulated, deafened guinea pig cochleae. *Laryngoscope* 97: 836-842.

[127] Luo, L., Koutnouyan, H., Baird, A., Ryan, A. F. (1993) Acidic and basic FGF mRNA expression in the adult and developing rat cochlea. *Hear. Res.* 69: 182-193.

[128] Malgrange, B., Rogister, B., Lefebvre, P. P., Mazy-Servais, C., Welcher, A. A., Bonnet, C., Hsu, R. Y., Rigo, J. M., Van De Water, T. R., Moonen, G. (1998) Expression of growth factors and their receptors in the postnatal rat cochlea. *Neurochem. Res.* 23: 1133-1138.

[129] Marin, V., Groveman, B. R., Qiao, H., Xu, J., Ali, M. K., Fang, X. Q., Lin, S. X., Rizkallah, R., Hurt, M. H., Bienkiewicz, E. A., Yu, X. M. (2010) Characterization of neuronal Src kinase purified from a bacterial expression system. *Protein Expr. Purif.* 74: 289-297.

[130] Martinez, R., Mathey-Prevot, B., Bernards, A., Baltimore, D. (1987) Neuronal $pp60^{c\text{-}src}$ contains a six-amino acid insertion relative to its non-neuronal counterpart. *Science* 237: 411-415.

[131] Matsubara, A., Laake, J. H., Davanger, S., Usami, S., Ottersen, O. P. (1996) Organization of AMPA receptor subunits at a glutamate synapse: a quantitative immunogold analysis of hair cell synapses in the rat organ of Corti. *J. Neurosci.* 16: 4457-4467.

[132] Mayer, M. L., Westbrook, G. L. (1987) The physiology of excitatory amino acids in the vertebrate central nervous system. *Prog. Neurobiol.* 28: 197-276.

[133] Mazzanti, M., DeFelice, L. J., Cohen, J., Malter, H. (1990) Ion channels in the nuclear envelope. *Nature* 343: 764-767.

[134] McBain, C. J., Mayer, M. L. (1994) N-Methyl-D-aspartic acid receptor structure and function. *Physiol. Rev.* 74: 723-760.

[135] McKenzie, G. J., Stevenson, P., Ward, G., Papadia, S., Bading, H., Chawla, S., Privalsky, M., Hardingham, G. E. (2005) Nuclear Ca2+ and CaM kinase IV specify hormonal- and Notch-responsiveness. *J. Neurochem.* 93: 171-185.

[136] McMahon, C. M., Patuzzi, R. B. (2002) The origin of the 900 Hz spectral peak in spontaneous and sound-evoked round-window electrical activity. *Hear. Res.* 173: 134-152.

[137] Mentzer, R. M., Jr., Lasley, R. D., Jessel, A., Karmazyn, M. (2003) Intracellular sodium hydrogen exchange inhibition and clinical myocardial protection. *Ann. Thorac. Surg.* 75: S700-S708.

[138] Miller, A. L., Prieskorn, D. M., Altschuler, R. A., Miller, J. M. (2003) Mechanism of electrical stimulation-induced neuroprotection: effects of verapamil on protection of primary auditory afferents. *Brain Res.* 966: 218-230.

[139] Mitchell, A., Miller, J. M., Finger, P. A., Heller, J. W., Raphael, Y., Altschuler, R. A. (1997) Effects of chronic high-rate electrical stimulation on the cochlea and eighth nerve in the deafened guinea pig. *Hear. Res.* 105: 30-43.

[140] Mizuno, M., Yamada, K., He, J., Nakajima, A., Nabeshima, T. (2003) Involvement of BDNF receptor TrkB in spatial memory formation. *Learn Mem.* 10: 108-115.

[141] Moolenaar, W. H., Defize, L. H., de Laat, S. W. (1986) Ionic signalling by growth factor receptors. *J. Exp. Biol.* 124:359-73: 359-373.

[142] Moolenaar, W. H., Tsien, R. Y., van der Saag, P. T., de Laat, S. W. (1983) Na+/H+ exchange and cytoplasmic pH in the action of growth factors in human fibroblasts. *Nature* 304: 645-648.

[143] Morton-Jones, R. T., Cannell, M. B., Housley, G. D. (2008) Ca2+ entry via AMPA-type glutamate receptors triggers Ca2+-induced Ca2+ release from ryanodine receptors in rat spiral ganglion neurons. *Cell Calcium* 43: 356-366.

[144] Mou, K., Hunsberger, C. L., Cleary, J. M., Davis, R. L. (1997) Synergistic effects of BDNF and NT-3 on postnatal spiral ganglion neurons. *J. Comp. Neurol.* 386: 529-539.

[145] Nagar, B., Hantschel, O., Seeliger, M., Davies, J. M., Weis, W. I., Superti-Furga, G., Kuriyan, J. (2006) Organization of the SH3-SH2 unit in active and inactive forms of the c-Abl tyrosine kinase. *Mol. Cell* 21: 787-798.

[146] Nicholls, D., Attwell, D. (1990) The release and uptake of excitatory amino acids. *Trends in Pharmacological Sciences* 11: 462-468.

[147] Nicolas-Puel, C., Faulconbridge, R. L., Guitton, M., Puel, J. L., Mondain, M., Uziel, A. (2002) Characteristics of tinnitus and etiology of associated hearing loss: a study of 123 patients. *Int. Tinnitus. J.* 8: 37-44.

[148] Niedzielski, A. S., Wenthold, R. J. (1995) Expression of AMPA, kainate, and NMDA receptor subunits in cochlear and vestibular ganglia. *J. Neurosci.* 15: 2338-2353.

[149] Nikolaeva, M. A., Mukherjee, B., Stys, P. K. (2005) Na+-dependent sources of intra-axonal Ca2+ release in rat optic nerve during in vitro chemical ischemia. *J. Neurosci.* 25: 9960-9967.

[150] Niu, X. W., Meech, R. W. (2000) Potassium inhibition of sodium-activated potassium (K(Na)) channels in guinea-pig ventricular myocytes. *J. Physiol.* 526 Pt 1: 81-90.

[151] Okada, M., Nada, S., Yamanashi, Y., Yamamoto, T., Nakagawa, H. (1991) CSK: a protein-tyrosine kinase involved in regulation of src family kinases. *J. Biol. Chem.* 266: 24249-24252.

[152] Okada, M., Nakagawa, H. (1989) A protein tyrosine kinase involved in regulation of pp60c-src function. *J. Biol. Chem.* 264: 20886-20893.

[153] Papadia, S., Stevenson, P., Hardingham, N. R., Bading, H., Hardingham, G. E. (2005) Nuclear Ca2+ and the cAMP response element-binding protein family mediate a late phase of activity-dependent neuroprotection. *J. Neurosci.* 25: 4279-4287.

[154] Parsons, S. J., Parsons, J. T. (2004) Src family kinases, key regulators of signal transduction. *Oncogene* 23: 7906-7909.

[155] Ponniah, S., Wang, D. Z., Lim, K. L., Pallen, C. J. (1999) Targeted disruption of the tyrosine phosphatase PTPalpha leads to constitutive downregulation of the kinases Src and Fyn. *Curr. Biol.* 9: 535-538.

[156] Puel, J. L. (1995) Chemical synaptic transmission in the cochlea. *Prog. Neurobiol.* 47: 449-476.

[157] Puel, J. L., Pujol, R., Tribillac, F., Ladrech, S., Eybalin, M. (1994) Excitatory amino acid antagonists protect cochlear auditory neurons from excitotoxicity. *J. Comp. Neurol.* 341: 241-256.

[158] Puel, J. L., Ruel, J., Gervais, d. C., Pujol, R. (1998) Excitotoxicity and repair of cochlear synapses after noise-trauma induced hearing loss. *Neuro Report* 9: 2109-2114.

[159] Ramekers, D., Versnel, H., Grolman, W., Klis, S. F. (2012) Neurotrophins and their role in the cochlea. *Hear. Res.* 288: 19-33.

[160] Repp, H., Birringer, J., Koschinski, A., Dreyer, F. (2001) Activation of a Ca2+-dependent K+ current in mouse fibroblasts by sphingosine-1-phosphate involves the protein tyrosine kinase c-Src. *Naunyn. Schmiedebergs. Arch. Pharmacol.* 363: 295-301.

[161] Roehm, P. C., Hansen, M. R. (2005) Strategies to preserve or regenerate spiral ganglion neurons. *Curr. Opin. Otolaryngol. Head Neck Surg.* 13: 294-300.

[162] Rogawski, M. A., Loscher, W. (2004) The neurobiology of antiepileptic drugs. *Nat. Rev. Neurosci.* 5: 553-564.

[163] Rose, C. R. (2002) Na+ signals at central synapses. *Neuroscientist.* 8: 532-539.

[164] Rose, C. R., Konnerth, A. (2001) NMDA receptor-mediated Na+ signals in spines and dendrites. *J. Neurosci.* 21: 4207-4214.

[165] Ruel, J., Chen, C., Pujol, R., Bobbin, R. P., Puel, J. L. (1999) AMPA-preferring glutamate receptors in cochlear physiology of adult guinea-pig. *J. Physiol.* 518 (Pt 3): 667-680.

[166] Ruel, J., Wang, J., Rebillard, G., Eybalin, M., Lloyd, R., Pujol, R., Puel, J. L. (2007) Physiology, pharmacology and plasticity at the inner hair cell synaptic complex. *Hear. Res* 227: 19-27.

[167] Rusznak, Z., Szucs, G. (2009) Spiral ganglion neurones: an overview of morphology, firing behaviour, ionic channels and function. *Pflugers. Arch.* 457: 1303-1325.

[168] Ryugo, D. K., Dodds, L. W., Benson, T. E., Kiang, N. Y. (1991) Unmyelinated axons of the auditory nerve in cats. *J. Comp. Neurol.* 308: 209-223.

[169] Salter, M. W., Kalia, L. V. (2004) Src kinases: a hub for NMDA receptor regulation. *Nat. Rev. Neurosci.* 5: 317-328.

[170] Salter, M. W., Pitcher, G. M. (2011) Dysregulated Src upregulation of NMDA receptor activity: a common link in chronic pain and schizophrenia. *FEBS J.* 279: 2-11.

[171] Scarpidis, U., Madnani, D., Shoemaker, C., Fletcher, C. H., Kojima, K., Eshraghi, A. A., Staecker, H., Lefebvre, P., Malgrange, B., Balkany, T. J., Van De Water, T. R. (2003) Arrest of apoptosis in auditory neurons: implications for sensorineural preservation in cochlear implantation. *Otol. Neurotol.* 24: 409-417.

[172] Schwartz, G., Fehlings, M. G. (2002) Secondary injury mechanisms of spinal cord trauma: a novel therapeutic approach for the management of secondary pathophysiology with the sodium channel blocker riluzole. *Prog. Brain Res.* 137: 177-190.

[173] Searchfield, G. D., Munoz, D. J., Thorne, P. R. (2004) Ensemble spontaneous activity in the guinea-pig cochlear nerve. *Hear. Res* 192: 23-35.

[174] Sheldon, C., Diarra, A., Cheng, Y. M., Church, J. (2004) Sodium influx pathways during and after anoxia in rat hippocampal neurons. *J. Neurosci.* 24: 11057-11069.

[175] Shepherd, R. K., Coco, A., Epp, S. B., Crook, J. M. (2005) Chronic depolarization enhances the trophic effects of brain-derived neurotrophic factor in rescuing auditory neurons following a sensorineural hearing loss. *J. Comp. Neurol.* 486: 145-158.

[176] Simmons, D. D., Liberman, M. C. (1988) Afferent innervation of outer hair cells in adult cats: I. Light microscopic analysis of fibers labeled with horseradish peroxidase. *J. Comp. Neurol.* 270: 132-144.

[177] Sin, W. C., Moniz, D. M., Ozog, M. A., Tyler, J. E., Numata, M., Church, J. (2009) Regulation of early neurite morphogenesis by the Na+/H+ exchanger NHE1. *J. Neurosci.* 29: 8946-8959.

[178] Sobko, A., Peretz, A., Attali, B. (1998) Constitutive activation of delayed-rectifier potassium channels by a src family tyrosine kinase in Schwann cells. *EMBO J.* 17: 4723-4734.

[179] Spoendlin, H. (1988) Neural anatomy of the ear. In *Physiology of the ear* (Janh, A. and Santos-Sacchi, J., eds) pp. 201-219, Raven, New York.

[180] Spoendlin, H., Brun, J. P. (1973) Relation of structural damage to exposure time and intensity in acoustic trauma. *Acta Otolaryngol.* 75: 220-226.

[181] Staecker, H., Kopke, R., Malgrange, B., Lefebvre, P., Van De Water, T. R. (1996) NT-3 and/or BDNF therapy prevents loss of auditory neurons following loss of hair cells. *Neuro Report* 7: 889-894.

[182] Strauss, O., Rosenthal, R., Dey, D., Beninde, J., Wollmann, G., Thieme, H., Wiederholt, M. (2002) Effects of protein kinase C on delayed rectifier K+ channel regulation by tyrosine kinase in rat retinal pigment epithelial cells. *Invest. Ophthalmol. Vis. Sci.* 43: 1645-1654.

[183] Strichartz, G., Rando, T., Wang, G. K. (1987) An integrated view of the molecular toxinology of sodium channel gating in excitable cells. *Annu. Rev. Neurosci.* 10:237-67: 237-267.

[184] Szabo, I., Gulbins, E., Apfel, H., Zhang, X., Barth, P., Busch, A. E., Schlottmann, K., Pongs, O., Lang, F. (1996) Tyrosine phosphorylation-dependent suppression of a voltage-gated K+ channel in T lymphocytes upon Fas stimulation. *J. Biol. Chem.* 271: 20465-20469.

[185] Teng, Y. D., Wrathall, J. R. (1997) Local blockade of sodium channels by tetrodotoxin ameliorates tissue loss and long-term functional deficits resulting from experimental spinal cord injury. *J. Neurosci.* 17: 4359-4366.

[186] Tezuka, T., Umemori, H., Akiyama, T., Nakanishi, S., Yamamoto, T. (1999) PSD-95 promotes Fyn-mediated tyrosine phosphorylation of the N-methyl-D- aspartate receptor subunit NR2A. *Proc. Natl. Acad. Sci. U.S.A.* 96: 435-440.

[187] Thiers, F. A., Nadol, J. B., Jr., Liberman, M. C. (2008) Reciprocal synapses between outer hair cells and their afferent terminals: evidence for a local neural network in the mammalian cochlea. *J. Assoc. Res. Otolaryngol.* 9: 477-489.

[188] Thomas, S. M., Brugge, J. S. (1997) Cellular functions regulated by Src family kinases. *Annu. Rev. Cell Dev. Biol.* 13: 513-609.

[189] Tiran, Z., Peretz, A., Sines, T., Shinder, V., Sap, J., Attali, B., Elson, A. (2006) Tyrosine phosphatases epsilon and alpha perform specific and overlapping functions in regulation of voltage-gated potassium channels in Schwann cells. *Mol. Biol. Cell* 17: 4330-4342.

[190] Toledo-Aral, J. J., Moss, B. L., He, Z. J., Koszowski, A. G., Whisenand, T., Levinson, S. R., Wolf, J. J., Silos-Santiago, I., Halegoua, S., Mandel, G. (1997) Identification of PN1, a predominant voltage-dependent sodium channel expressed principally in peripheral neurons. *Proc. Natl. Acad. Sci. U S A* 94: 1527-1532.

[191] Tong, Q., Stockand, J. D. (2005) Receptor tyrosine kinases mediate epithelial Na(+) channel inhibition by epidermal growth factor. *Am. J. Physiol. Renal. Physiol.* 288: F150-F161.

[192] Trellakis, S., Lautermann, J., Lehnerdt, G. (2007) Lidocaine: neurobiological targets and effects on the auditory system. *Prog. Brain Res.* 166: 303-322.

[193] Trepanier, C. H., Jackson, M. F., MacDonald, J. F. (2011) Regulation of NMDA receptors by the tyrosine kinase Fyn. *FEBS J.* 279: 12-19.

[194] Trump, B. F., Berezesky, I. K. (1995) Calcium-mediated cell injury and cell death. *FASEB J.* 9: 219-228.

[195] van Loon, M. C., Ramekers, D., Agterberg, M. J., de Groot, J. C., Grolman, W., Klis, S. F., Versnel, H. (2013) Spiral ganglion cell morphology in guinea pigs after deafening and neurotrophic treatment. *Hear. Res* 298: 17-26.

[196] Verkhratsky, A. J., Petersen, O. H. (1998) Neuronal calcium stores. *Cell Calcium* 24: 333-343.

[197] Vornov, J. J., Thomas, A. G., Jo, D. (1996) Protective effects of extracellular acidosis and blockade of sodium/hydrogen ion exchange during recovery from metabolic inhibition in neuronal tissue culture. *J. Neurochem.* 67: 2379-2389.

[198] Wang, Q., Green, S. H. (2011) Functional role of neurotrophin-3 in synapse regeneration by spiral ganglion neurons on inner hair cells after excitotoxic trauma in vitro. *J. Neurosci.* 31: 7938-7949.

[199] Waxman, S. G., Hains, B. C. (2006) Fire and phantoms after spinal cord injury: Na+ channels and central pain. *Trends Neurosci.* 29: 207-215.

[200] Weber, J. T. (2012) Altered calcium signaling following traumatic brain injury. *Front Pharmacol.* 3: 60.

[201] Werling, L. L., Brown, S. R., Puttfarcken, P., Cox, B. M. (1986) Sodium regulation of agonist binding at opioid receptors. II. Effects of sodium

replacement on opioid binding in guinea pig cortical membranes. *Mol. Pharmacol.* 30: 90-95.

[202] Xin, W. K., Kwan, C. L., Zhao, X. H., Xu, J., Ellen, R. P., McCulloch, C. A., Yu, X. M. (2005) A functional interaction of sodium and calcium in the regulation of NMDA receptor activity by remote NMDA receptors. *J. Neurosci.* 25: 139-148.

[203] Xin, W. K., Zhao, X. H., Xu, J., Lei, G., Kwan, C. L., Zhu, K. M., Cho, J. S., Duff, M., Ellen, R. P., McCulloch, C. A., Yu, X. M. (2005) The removal of extracellular calcium: a novel mechanism underlying the recruitment of N-methyl-d-aspartate (NMDA) receptors in neurotoxicity. *Eur. J. Neurosci.* 21: 622-636.

[204] Xu, J., Weerapura, M., Ali, M. K., Jackson, M. F., Li, H., Lei, G., Xue, S., Kwan, C. L., Manolson, M. F., Yang, K., MacDonald, J. F., Yu, X. M. (2008) Control of excitatory synaptic transmission by C-terminal Src kinase. *J. Biol. Chem.* 283: 17503-17514.

[205] Yamada, K., Nabeshima, T. (2003) Brain-derived neurotrophic factor/TrkB signaling in memory processes. *J. Pharmacol. Sci.* 91: 267-270.

[206] Yang, T., Kersigo, J., Jahan, I., Pan, N., Fritzsch, B. (2011) The molecular basis of making spiral ganglion neurons and connecting them to hair cells of the organ of Corti. *Hear. Res.* 278: 21-33.

[207] Yu, X. M. (2006) The role of intracellular sodium (Na^+) in the regulation of NMDA receptor-mediated channel activity and toxicity. *Mol. Neurobiol.* 3: 63-79.

[208] Yu, X. M., Askalan, R., Keil, G. J. I., Salter, M. W. (1997) NMDA channel regulation by channel-associated protein tyrosine kinase Src. *Science* 275: 674-678.

[209] Yu, X. M., Groveman, B. R. (2012) Src family kinases in the nervous system. *FEBS J.* 279: 1.

[210] Yu, X. M., Salter, M. W. (1998) Gain control of NMDA-receptor currents by intracellular sodium. *Nature* 396: 469-474.

[211] Yu, X. M., Salter, M. W. (1999) Src, a molecular switch governing gain control of synaptic transmission mediated by N-methyl-D-aspartate receptors. *Proc. Natl. Acad. Sci. U.S.A.* 96: 7697-7704.

[212] Yuan, A., Santi, C. M., Wei, A., Wang, Z. W., Pollak, K., Nonet, M., Kaczmarek, L., Crowder, C. M., Salkoff, L. (2003) The sodium-activated potassium channel is encoded by a member of the Slo gene family. *Neuron* 37: 765-773.

[213] Zeng, C., Yang, Z., Shreve, L., Bledsoe, S., Shore, S. (2012) Somatosensory projections to cochlear nucleus are upregulated after unilateral deafness. *J. Neurosci.* 32: 15791-15801.

[214] Zhang, D., Hou, Q., Wang, M., Lin, A., Jarzylo, L., Navis, A., Raissi, A., Liu, F., Man, H. Y. (2009) Na,K-ATPase activity regulates AMPA receptor turnover through proteasome-mediated proteolysis. *J. Neurosci.* 29: 4498-4511.

[215] Zhang, Z., Fan, J., Ren, Y., Zhou, W., Yin, G. (2013) The release of glutamate from cortical neurons regulated by BDNF via the TrkB/Src/PLC-gamma1 pathway. *J. Cell Biochem.* 114: 144-151.

[216] Zhong, S. X., Liu, Z. H. (2004) Immunohistochemical localization of the epithelial sodium channel in the rat inner ear. *Hear. Res.* 193: 1-8.

[217] Zipfel, G. J., Babcock, D. J., Lee, J. M., Choi, D. W. (2000) Neuronal apoptosis after CNS injury: the roles of glutamate and calcium. *J. Neurotrauma.* 17: 857-869.

Index

A

B

G

H

I

Q

R

S

T

U

V

W

Y